Functional Foods

The Consumer, the Products and the Evidence

Functional Foods
The Consumer, the Products and the Evidence

Edited by

Michèle J. Sadler
The British Nutrition Foundation, London, UK

Michael Saltmarsh
Inglehurst Foods, Alton, Hampshire, UK

THE ROYAL
SOCIETY OF
CHEMISTRY
Information
Services

Proceedings of a joint conference held by the British Nutrition Foundation and the Food Chemistry Group of the Royal Society of Chemistry on 2-4 April 1997 at Wye College, University of London, Kent, UK.

Special Publication No. 215

ISBN 0-85404-792-1

A catalogue record for this book is available from the British Library

Published by The Royal Society of Chemistry,
Thomas Graham House, Science Park, Milton Road,
Cambridge CB4 4WF, UK

Printed by Bookcraft (Bath) Ltd.

Foreword

M. Spiro

Immediate Past Chairman
Royal Society of Chemistry
Food Chemistry Group
Burlington House
Piccadilly, London
W1V 0BN

It gives me great pleasure to introduce these Proceedings based on the conference on Functional Foods 97 held at Wye College from 2-4 April 1997, and organised jointly by the British Nutrition Foundation and the Food Chemistry Group of the Royal Society of Chemistry.

We are delighted that delegates were present from so many different backgrounds: from food manufacturing and processing industries, food retailing firms, universities, research and government institutes, and from such a large number of overseas countries including Ireland, Finland, Denmark, the Netherlands, the USA, Germany, Belgium, France and Spain as well as from almost all parts of Britain. This demonstrates how wide is the interest in this subject and how topical it has become.

If we look back at recent history, it would be fair to say that in the 1960s the main preoccupation seemed to be with sex, in the 1970s with drugs, in the 1980s (particularly in Mrs Thatcher's Britain) with money, while in the 1990s a major concern is with food and health. News items about these subjects abound: only a week before the conference major heart charities emphasized the importance of eating enough fruit and vegetables, clearly functional foods in their own right. A good indicator of the current public interest in food is British television since people in the United Kingdom are said to watch more TV than those in other European countries. Hardly an evening goes by without a programme on this subject on the four main terrestrial channels - I can recall at least seven recent and current TV series on food, three on drink and four competition or other entertainment programmes based on food and drink. For example, on the night before the conference several delegates will no doubt have watched the regular BBC *Food and Drink* show, while during the conference they missed seeing *Ready Steady Cook*, a food-based episode of *An Inspector Calls*, *Ken Hom's Hot Wok*, and *A Cook on the Wild Side*. However, the organisers of the conference thoughtfully provided even more interesting live food and drink entertainment in the evenings, including an English Wine Tasting and a Conference Banquet.

The media have clearly recognised that food is news and, as always, that health is news. Functional foods encompass both food and health, which explained the presence of TV cameras in the hall at the start of the conference as well as a radio interview behind the scenes. Government committees too, have at last begun to appreciate the importance of the subject and the UK Technology Foresight Panel on Food and Drink has identified diet and health as major areas for future research.

The format of Functional Foods 97 allowed for much frank and open discussion, not only during the formal sessions but throughout the whole meeting. The conference thus provided a unique opportunity for those concerned with different aspects of the subject to meet together and to understand better the problems, solutions, and successes in all the relevant areas, and what the future is likely to bring.

I believe that Functional Foods 97 has indeed met its objectives of enabling delegates from diverse backgrounds to stay at the forefront of developments in this rapidly growing field and to forge useful links with other parties involved. It is much to the credit of the organisers that the meeting also proved so enjoyable and was set at such a pleasant location.

Editors' Introduction

These are the proceedings of a joint conference held by the British Nutrition Foundation and the Food Chemistry Group of the Royal Society of Chemistry on 2-4 April 1997 at Wye College, University of London, in Kent.

The food industry's recent focus on developing products with positive nutritional benefits has occurred in the light of the findings of the UK Technology Foresight Panel on Food and Drink, in which diet and health was identified as one of the main drivers of future innovation. Thus we felt it timely to hold a conference to critically review these developments.

The aims and scope of the conference were to bring together academic research, product development and market research to critically review three areas concerning the potential role of functional foods in improving human health. The three key areas were:

- the evidence for benefit of physiologically functional ingredients for human health
- the technological challenges of incorporating such ingredients into products in physiologically relevant amounts
- consumer and regulatory issues.

The programme included six plenary lectures reviewing these three areas, and contributions in the form of presentations and posters, which provided a stimulus to fully debate the issues, in the light of the most recent data.

We are grateful for all the scientific presentations which helped to make the conference such a success. We are also grateful for financial support received from Unilever, United Biscuits, BIBRA International, Eridania Béghin-Say, The Rowett Research Institute, and TNO Nutrition and Food Research Institute. Finally we would like to thank Elisa Pons of the British Nutrition Foundation for ensuring that the conference ran smoothly and for her valuable help with the production of the conference proceedings.

We hope that publication of these proceedings will enable a wider audience to share the scientific content of the meeting, and that this volume will be a valuable contribution to the area.

Michèle J Sadler and
Michael Saltmarsh

Contents

III Technological Aspects

IV Regulatory and Consumer Issues

Contributors

H Adlercreutz
Department of Clinical Chemistry, University of Helsinki, Meilahti Hospital, FIN-00290 Helsinki, Finland

CM Aguilera
Department of Biochemistry and Molecular Biology, Institute of Nutrition and Food Technology, University of Granada, Granada, Spain

M Alander
VTT Biotechnology and Food Research, Espoo, Finland

AJ Alldrick
Cereals & Cereals Processing Division, Campden & Chorleywood Food Research Association, Chipping Campden, GL55 6LD

A-M Aura
VTT Biotechnology and Food Research, Espoo, Finland

PJ Beers
Food Research Centre, University of Lincoln Grimsby Campus, 61 Bargate, Grimsby, DN34 5AA

Professor A Bender
2 Willow Vale, Fetcham, Leatherhead, Surrey, TK22 9TE

IFF Benzie
Department of Health Sciences, The Hong Kong Polytechnic University, Kowloon , Hong Kong

N Bissetti
UESPD, Bât 440, INRA, 78352 Jouy-en-Josas Cedex, France

E Bontebal
PURAC biochem, PO Box 21, 4200 AA Gorichem, The Netherlands

EA Bowey
Department of Microbiology and Nutrition, BIBRA International, Carshalton, Surrey, SM5 4DS

G Cardeni
Department of Pharmacology, University of Florence, 50134 Florence, Italy

MA Carrión-Gutiérrez
ASAC Medical Department, Alicante, Spain

J Coutts
BIBRA International, Carshalton, Surrey, SM5 4DS

A Cresci
University of Camerino, 62032 Camerino, Italy

N Delzenne
Unité de Biochimie Toxicologique et Cancérogique, Département des sciences pharmaceutiques, Université Catholique de Louvain, UCL-BCTC 7369, 73 Avenue Mounier, B-1200, Belgium

P Dolara
Department of Pharmacology, University of Florence, 50134 Florence, Italy

C Dubuquoy
UESPD, Bât 440, INRA, 78352 Jouy-en-Josas Cedex, France

PR Ellis
Biopolymers Group, King's College London, London W8 7AH

M Fabritius
VTT Biotechnology and Food Research, Espoo, Finland

R Fuller
Russet House, Ryeish Green, Reading, RG6 6BZ

AP Garnett
Nutrition, Food and Health Research Centre, King's College London, Campden Hill Road, London, W8 7AH

GR Gibson
Microbiology Department, Institute of Food Research, Reading, RG6 6BZ

Professor A Gil
Department of Biochemistry and Molecular Biology, Institute of Nutrition and Food Technology, University of Granada, Granada, Spain

Professor JMT Hamilton-Miller
Department of Medical Microbiology, Royal Free Hospital School of Medicine, London, NW3 2QG

W-C Huang
Nutrition, Food and Health Research Centre, King's College London, Campden Hill Road, London, W8 7AH

SJ Hurley
Sheffield University, Department of Surgical & Anaesthetic Sciences - K Floor, The Royal Hallamshire Hospital, Sheffield, S10 2JF

E Isolauri
University of Turku, Finland

N Kok
Unité de Biochimie Toxicologique et Cancérogique, Département des sciences pharmaceutiques, Université Catholique de Louvain, UCL-BCTC 7369, 73 Avenue Mounier, B-1200, Belgium

TP Kravtchenko
Colloides Naturels International, 129 Chemin de Croisset, BP 4151, F-76723 Rouen, Cedex, France

L Lievense
Unilever Research Vlaardingen, Olivier van Noortlaan 120, 3133 AT Vlaardingen, The Netherlands

P Lim
Nutrition, Food and Health Research Centre, King's College London, Campden Hill Road, London, W8 7AH

S Madsen
MD Foods amba, Skanderborgvej 277, PO Box 2470, DK-8260 Viby J, Denmark

GT McAnlis
Department of Food Science, The Queens University of Belfast, Newforge Lane, Belfast, BT9 5PX
Department of Clinical Biochemistry, The Queens University of Belfast, Royal Victoria Hospital, Belfast, BT12 6BJ

J McEneny
Department of Clinical Biochemistry, The Queens University of Belfast, Royal Victoria Hospital, Belfast, BT12 6BJ

MC Moreau
UESPD, Bât 440, INRA, 78352 Jouy-en-Josas Cedex, France

LM Morgan
Surrey University, Department of Biological Sciences, Guildford

G Morozzi
University of Perugia, Perugia, Italy

AEJ Morris
Food Research Centre, University of Lincoln Grimsby Campus, 61 Bargate, Grimsby, DN34 5AA

M Ni Néill
Department of Biological Sciences, Dublin Institute of Technology, Dublin 8, Ireland

JD O'Reilly
Nutrition, Food and Health Research Centre, King's College London, Campden Hill Road, London, W8 7AH

JF Payne
Food Research Centre, University of Lincoln Grimsby Campus, 61 Bargate, Grimsby, DN34 5AA

J Pearce
Department of Food Science, The Queens University of Belfast, Newforge Lane, Belfast, BT9 5PX

B Pool-Zobel
Bundesforschungsanstalt für Ernährung, 76131 Karlsruhe, Germany

RE Potjewijd
PURAC biochem, PO Box 21, 4200 AA Gorinchem, The Netherlands

K Poutanen
VTT Biotechnology and Food Research, Espoo, Finland

A Ramírez-Boscá
ASAC Medical Department, Alicante, Spain

MC Ramírez-Tortosa
Department of Biochemistry and Molecular Biology, Institute of Nutrition and Food Technology, University of Granada, Granada, Spain

M Rayner
British Heart Foundation Health Promotion Research Group, Division of Public Health and Primary Health Care, University of Oxford, Oxford, OX2 6HE

Professor DP Richardson
Nestlé UK Ltd, St George's House, Croydon, Surrey, CR9 1NR

TJ Ridgway
Department of Applied Biochemistry and Food Science, University of Nottingham, Sutton Bonington Campus, Loughborough, Leceistershire, LE12 5RD

Professor M Roberfroid
Unité de Biochimie Toxicologique et Cancérogique, Département des sciences pharmaceutiques, Université Catholique de Louvain, UCL-BCTC 7369, 73 Avenue Mounier, B-1200, Belgium

S Ross-Murphy
Biopolymers Group, King's College London, London W8 7AH

Professor IR Rowland
Human Nutrition Research Group, University of Ulster, Coleraine, BT52 1SA

C Rumney
BIBRA International, Carshalton, Surrey, SM5 4DS

MJ Sadler *
The British Nutrition Foundation, High Holborn House, 52-54 High Holborn, London WC1V 6RQ

E Salminen
University of Turku, Finland

A Salter
Department of Applied Biochemistry and Food Science, University of Nottingham, Sutton Bonington Campus, Loughborough, Leicestershire, LE12 5RD

M Saltmarsh
Inglehurst Foods, 53 Blackberry Lane, Four Marks, Alton, Hants, GU34 5DF

Professor TAB Sanders
Nutrition, Food and Health Research Centre, King's College London, Campden Hill Road, London, W8 7AH

M Saxelin
Valio R&D, Helsinki

VA Sessions
Department of Applied Biochemistry and Food Science, University of Nottingham, Sutton Bonnington Campus, Loughborough, Leicestershire, LE12 5RD

S Shah
Department of Medical Microbiology, Royal Free Hospital School of Medicine, London, NW3 2QG

C Shortt
Yakult UK Ltd, 12-16 Telford Way, Westway Estate, Acton, London, W3 7XS

S Silvi
Universty of Camerino, 62032 Camerino, Italy

Professor M Spiro
Royal Society of Chemistry, Food Chemistry Group, Burlington House, Piccadilly, London, W1V 0BN

Professor JJ Strain
Human Nutrition Research Group, University of Ulster, Coleraine, BT52 1SA

*Present address: Institute of Grocery Distribution, Letchmore Heath, Watford WD2 8DQ, UK.

T Suorti
VTT Biotechnology and Food Research, Espoo, Finland

J Tomlin
Sheffield University, Department of Surgical & Anaesthetic Sciences - K Floor, The Royal Hallamshire Hospital, Sheffield, S10 2JF

GA Tucker
Department of Applied Biochemistry and Food Science, University of Nottingham, Sutton Bonington Campus, Loughborough, Leceistershire, LE12 5RD

G van Poppel
TNO Nutrition and Food Research Institute, Utrechtseweg 48, PO Box 360, 3700 AJ Zeist, The Netherlands

C van Aarle
PURAC biochem, PO Box 21, 4200 AA Gorinchem, The Netherlands

H Verhagen
TNO Nutrition and Food Research Institute, Utrechtseweg 48, PO Box 360, 3700 AJ Zeist, The Netherlands

A von Wright
VTT Biotechnology and Food Research, Espoo, Finland

RW Welch
Northern Ireland Research Centre for Diet and Health, University of Ulster, Coleraine, BT52 1SA

J Winkler
28 St Paul Street, London, N1 7AB.

H Wiseman
Nutrition, Food and Health Research Centre, King's College London, Campden Hill Road, London, W8 7AH

IS Young
Department of Clinical Biochemistry, The Queens University of Belfast, Royal Victoria Hospital, Belfast, BT12 6BJ

J Young
Leatherhead Food RA, Randalls Road, Leatherhead, Surrey, KT22 7RY

KM Younger
Department of Biological Sciences, Dublin Institute of Technology, Dublin 8, Ireland

I Pro- and Prebiotics

THE ROLE OF PROBIOTICS AND PREBIOTICS IN THE FUNCTIONAL FOOD CONCEPT

G.R. GIBSON[1] and R. FULLER[2]

[1] Microbiology Department, Institute of Food Research, Reading RG6 6BZ
[2] Russet House, Ryeish Green, Reading RG7 1ES

1 INTRODUCTION

Consumers are increasingly exposed to foods touted as 'functional.' The concept has an enormous potential market value (estimated at up to £60 billion per annum), yet the definition and role of these foodstuffs remains unclear–attracting much derision from some consumer agencies and other 'experts.'

A functional food has many possible definitions, not least of which is the reality that all foods have at least one function - to help prevent starvation! However, a functional food (or nutraceutical, vitafood, pharmafood, designer food, food for specified health use) should be considered as a dietary ingredient that may have health attributes over and above its normal nutritional value. More formally, one usable definition of a functional food is '*a dietary ingredient that affects its host in a targeted manner so as to exert positive effects that may, in due course, justify certain health claims*'.[1] Whilst this is an appropriate definition, the question remains as to how foodstuffs can be made more 'functional.' There are four conceivable mechanisms.[2]

i)   elimination of a component that has a negative physiological effect, eg allergenic, toxic or mutagenic compounds
ii) increased concentration of a component that may have a beneficial aspect, eg dietary fibre
iii) addition of a novel ingredient seen as advantageous, eg a vitamin and/or mineral
iv) partial replacement of a negative component, with a benign or positive entity that does not adversely affect nutritive value, eg fat substitution with certain long chain carbohydrates.

If the concept is to gain improved scientific credibility and lead to proven health advantages, it is important that all possible aspects of ingestion are considered–with carefully conducted and controlled volunteer trials being carried out. The role of the human gastrointestinal microflora in host health and nutrition is critical.

Because of its resident bacterial flora, the colon is the most metabolically active organ in the human body. Whilst the large intestine is involved in the aetiology and maintenance of many disease states, the microbiota is also critical for host wellbeing. Foodstuffs provide the principal growth substrates for colonic bacteria. Therefore, it is rational to conceive a functional food approach for the gut microflora.

To generalise, it is possible to categorise colonic bacterial components on the basis of whether they exert potentially pathogenic or health promoting aspects. For obvious reasons, there is interest in increasing the numbers and activities of the latter in the large gut. Clearly therefore, dietary influences on the large gut microbiota composition and activities have a major role in the functional food concept. One approach for improved microflora management involves live microorganisms as probiotics (usually lactobacilli or bifidobacteria). The use of probiotics is now widespread and includes the supplementation of fermented milk products as well as 'over the counter' freeze dried preparations.

Prebiotics are dietary components that have a specific fermentation directed towards certain populations of indigenous gut bacteria. Some oligosaccharides have the potential to stimulate bifidobacteria in the colon to such an extent that, after a short feeding period, they become numerically predominant in faeces.

The clinical and health promoting aspects of food supplementation with probiotics and prebiotics are currently being defined. However, it would appear that this technology is critical for product developments that influence large gut bacteriology and, possibly, the outcome of certain gastrointestinal diseases. The advancement of molecular biological methodologies, that more accurately determine the flora composition and activities in response to diet, is likely to accelerate this potential.

2 PROBIOTICS

A probiotic is '*a live microbial feed supplement which beneficially affects the host animal by improving its intestinal microbial balance*'.[3,4] For humans, this definition mainly impacts on the complex colonic microbiota. As indicated earlier, the probiotic organism(s) used for human consumption is usually a lactic acid excretor - although the definition opens up the possibility of colonisation by any fed microorganism that has the ability to exert a beneficial effect. This would usually require some degree of *in situ* colonisation.

The history of probiotics dates back as far as the first intake of fermented milks, some 2000 years ago. However, it is in probably the work of Metchnikoff[5] in the early years of this century, that the first scientific assessments of probiotics were made. Metchnikoff hypothesised that beneficial effects of the consumption of soured milks was possibly because of a direct antagonistic effect on pathogenic bacteria in the hindgut. The begining of the probiotic concept was probably Metchnikoff's work with *Lactobacillus delbrueckii* subsp. *bulgaricus* and *Streptococcus salivarius* subsp. *thermophilus*. This formulation still provides the basis of classical yoghurt manufacturing. Subsequently, *Lactobacillus acidophilus*, a human derived strain, was used in fermented milk preparations. Today, many different microorganisms are added to yoghurts for their probiotic potential. These include a) Lactobacilli such as *L.acidophilus*, *L.casei*, *L.delbrueckii* subsp. *bulgaricus, L.reuteri*, *L.brevis*, *L.cellobiosus*, *L.curvatus*, *L.fermentum*, *L.plantarum;* b) Gram positive cocci such as *Lactococcus lactis* subsp. *cremoris, Streptococcus salivarius* subsp. *thermophilus, Enterococcus faecium, S.diaacetylactis, S.intermedius* and c) Bifidobacteria such as

B.bifidum, B.adolescentis, B.animalis, B.infantis, B.longum , B.thermophilum. Much of the pioneering work with bifidobacteria was carried out by Tissier.[6]

Care should be taken in selecting appropriate probiotics. Huis in't Veld and Shortt[7] have suggested three general areas for good selection criteria. These are shown in Table 1. These criteria require a combination of *in vitro* as well as *in vivo* studies.

Three important pieces of work contribute towards the development of probiotics as possible protective factors in the gastrointestinal tract. Firstly, Bohnoff *et al.* [8] showed that antibiotic treated mice had a greater susceptibility to infection than untreated mice. The implication was that suppression of the normal gut flora removed colonisation resistance to the pathogen. Secondly, studies with germfree and conventional animals demonstrated that the sterile forms had a far higher susceptibility to infection.[9] Finally, it is known that fecal enemas can influence the survival of gut pathogens. In this context, the toxic nature of *Clostridium difficile*, which occurs in association with antibiotic therapy, can be alleviated using fecal preparations.[10] Obviously, this form of gut therapy has many aesthetical problems, so the development of pure cultures which can be ingested orally has been prudent. These may take the form of powders, pastes, capsules, liquid and semi-solid suspensions, tablets and sprays. Probiotics are widely used in agriculture and are now also common products for human consumption. They are thought to exert their effect by modification of the intestinal microflora.

The extent by which a probiotic is able to exert microbiota alterations can be difficult to assess. Measurement of changes in viable numbers in feces after ingestion can be difficult to interpret as complexities occur in separating the fed strains from those that are indigenous. For an improved efficacy, genetic techniques are required.

Table 1 *Selection Criteria for Probiotics (modified from reference 7)*

a) General aspects
- origin of the strain
- safety
- survivability in the production and after ingestion (eg resistance to acid, pancreatic secretions and bile)
- viability
b) Technological aspects
- production characteristics
- processing
- sensory properties
c) Functional aspects
- microbiological properties
- effects on the consumer
- adherence
- effects on pathogens
- modulation of metabolic activities
- immunomodulatory effects

Hitherto used traditional gut microbiological procedures that rely on phenotypic differences between organisms such as different morphologies, biochemical tests, etc. are probably unreliable, as the bacteria may exhibit some metabolic plasticity. Moreover, the techniques are prone to operator subjectivity (eg in the interpretation of Analytical Profile Index results). The full microbial diversity of the gut flora remains undescribed and current methodologies are too laborious, time consuming and expensive to develop much further–and certainly to the required extent. Molecular techniques offer an attractive solution to this problem. For example, molecular sequence analyses, particularly of rRNA, provide a robust tool for determining the genetic interrelationships of microorganisms. Genetic fingerprinting offers an extremely reliable and sensitive means of determinative microbiology, at the species and sub-species levels, with non-culturable microorganisms also being accessible to high fidelity characterisation and identification. By utilising diagnostic sequences within the rRNA it is also possible to design gene probes for the unequivocal detection of microorganisms (eg probiotics) in a mixed culture environment. The approach has recently been used for probiotic bifidobacteria administered to infants.[11]

Another alternative by which probiotic organisms can be demonstrated to pass through the gastrointestinal tract is by the use of specific biomarkers. This has been carried out with selected antibiotic resistant strains. However, it may be that *in situ* transfer of plasmid DNA may occur, again giving a degree of uncertainty to the results. A future approach may involve the tagging of probiotic strains with conserved genetic markers such as those coding for fluorescent proteins or specific detectable oligonucleotides.

Probiotic effects are probably influenced very strongly by the ability of the organism to survive and multiply within the host. The probiotic should be metabolically stable and active. In humans, the study material is almost always fecal specimens. As such it is not possible to detect changes in the proximal colon, or assess the degree of colonisation in more anterior regions of the gastrointestinal tract. Advanced *in vitro* modelling systems may be useful to predict bacteriological changes, but are not a substitute for the *in vivo* situation. Animal models are preferred by some researchers, but there are differences between the human bacterial profile and those of other mammals. Pigs are thought to give the closest analogy. Other work has concentrated on the use of germ free rodents inoculated with human bacteria. These give a reasonable approximation, but it is unclear how long the stability of the inoculum profile is maintained, whilst handling procedures prior to inoculation may confer added variability. The interpretation of data should always be carried out in light of the possible limitations on the type of model system used. Nevertheless, some interesting results have arisen on the possible therapeutic value of probiotics (see later).

3 PREBIOTICS

One of the most critical aspects of probiotics is that they must remain viable after ingestion. It is possible to select strains that are able to survive well when exposed to gastric acid or certain small intestinal secretions. In this context, it has been shown by aspiration of the terminal ileum of human volunteers fed bifidobacteria, that probiotics are able to enter the colon.[12] However, the large intestine is occupied by a very diverse and metabolically active microbiota. It may be that probiotic survival is difficult in such a situation, particularly as the live feed addition may have been compromised in the upper gastrointestinal tract. Both lactobacilli[13] and bifidobacteria[14] disappear from faeces rapidly after probiotic feeding, indicating that long term colonisation does not occur. A possible alternative (or addition) is the use of prebiotics.

A prebiotic is *'a non-digestible food ingredient that beneficially affects the host by selectively stimulating the growth and/or activity of one or a limited number of bacteria in the colon, that can improve host health'*.[1] The key difference between this concept and probiotics is the use of non-viable food ingredients. Any dietary component that is unabsorbed through passage to the colon could possibly act as a prebiotic. However, to be effective, the prebiotic must have a selective fermentation such that the composition of the large intestinal microbiota is altered towards a potentially healthier community. The concept is therefore built around the fact that the normal gut flora has both pathogenic and beneficial entities. For example, the end products of colonic bacterial fermentation have varying effects on host health. Short chain fatty acids may be absorbed for increased energy gain, whilst certain bacterial species may help in gas distension problems. On the contrary, the accumulation of proteolytic end products such as ammonia, amines and phenols may have potentially deleterious effects. Table 2 gives examples of both health promoting and harmful aspects of human gut microbiology. An improved colonic flora composition may be achieved through selectively targeting the metabolism of remedial genera like lactobacilli or bifidobacteria, although a suppression of virulence in pathogenic microorganisms is another approach. Because of their limted metabolism, certain oligosaccharides are primary candidates. Most work, particularly by Japanese researchers, has been carried out with those that contain fructose.[15-18]

3.1 Fructooligosaccharides as Prebiotics

Fructo-oligosaccharides are short and medium length chains of β-D fructans in which fructosyl units are bound by a β 2-1 osidic linkage. Their synthesis in plant cells starts by the transfer of a fructosyl moiety between two sucrose molecules,[19] some of these molecules have a glucose unit as the initial moeity. The β 2-1 osidic bond of fructo-oligosaccharides is not hydrolysed by mammalian digestive enzymes,[20] but is susceptible to attack by certain bacteria. Inulin has a large degree of polymerisation of up to 60 and is the energy reserve in well over 30,000 plants. Both linear and branched chain varieties of the molecule exist (Figure 1). Predominant sources of fructooligosaccharides in the western diet include wheat, onion, garlic, asparagus, chicory, banana and artichoke.

Using data from *in vitro* studies that indicated the specific fermentation that fructooligosaccharides have,[21] a human volunteer trial was instigated to assess the bifidogenic effect of fructooligosaccharides *in vivo*. In the latter experiments, the influence of this carbohydrate on the faecal bacterial composition in healthy persons was evaluated during a 45-day feeding period, when the volunteers were given a strictly controlled diet.[22] Eight volunteers participated in the oligofructose feeding experiment. They had never suffered from any form of gastrointestinal disorder and had not taken antibiotics for at least three months before the start of the study. Energy requirements for each volunteer during the feeding regime were calculated on the basis of body weight. During the first five days, subjects were given a non-controlled diet. During this time, a stool sample was collected for bacteriological analysis. Subsequently, subjects

Table 2 *Examples of Potentially Harmful and Health Promoting Aspects of the Human Large Intestinal Microbiota*

Pathogenic	Beneficial
Intestinal putrefaction	Maintenance of homeostasis
Tissue invasion	Production of vitamins
Potentially carcinogenic	Metabolism of procarcinogens
Toxin production	Stimulation of immunity
Cytotoxicity	Improved energy yield
Diarrhoea/constipation	Lower gas distension
Inflammatory bowel disease	Production of butyrate and other short chain fatty acids
Site of gut infections	Inhibition of invading species
Liver damage	Metabolism of xenobiotic compounds
Antibiotic associated disease	Reduction of translocation

were given the controlled diet supplemented with 15 g of sucrose for a 15 day period. This was then replaced by fructooligosaccharides for a further 15 days, followed by another period with sucrose. Stool samples were taken periodically for bacterial enumeration. The use of fructooligosaccharides as a replacement for sucrose in the diet caused a marked increase in bifidobacteria, whilst bacteroides, fusobacteria and clostridia all decreased. Other bacteria counted (total aerobes, total anaerobes, lactobacilli, coliforms and Gram positive cocci) remained more or less unchanged. As such, the selectivity of the oligosaccharides was demonstrated. Similar data were obtained with inulin.[22] It is inadequate to classify a substrate as a prebiotic because it is fermented by lactobacilli and/or bifidobacteria–effects on other bacteria are equally important and need to be taken into account, ie the prebiotic should not be generally fermented. The prebiotic concept is shown schematically in Figure 2.

3.2 Other Oligosaccharides as Prebiotics

Crittenden and Playne [23] have discussed the various production and properties of certain food grade oligosaccharides. Apart from fructooligosaccharides, galactooligosaccharides appear to be good prebiotics. Their feeding to human flora associated rats has the effect of increasing numbers of bifidobacteria and lactobacilli, while decreasing enterobacteria.[24] Other prebiotic oligosaccharides may include raffinose, palatinose and stachyose, as well as those that contain xylose, maltose, mannose, and soya. However, their specific activity towards an improved colonic health requires confirmation, ie the selectivity of the fermentation, as well as degree of absorption in the upper gastrointestinal tract may be questionable. The disaccharide lactulose is able to target the colon, is well fermented therein and has been used to increase lactobacilli in the intestine of bottle-fed infants.[25]

The area requires further clarification, as it may be that other oligosaccharides may exert more preferential effects than those that contain fructose. The design of novel oligosaccharides, that may have improved sensory and biological properties is also worthy of further investigation.

4 HEALTH ASPECTS OF GUT FLORA MANIPULATION

It is important that the possible health advantages of probiotics and prebiotics are assessed using valid scientific approaches and, preferably, involving human volunteer trials. A number of areas may arouse interest, but the evidence for an involvement of gut flora manipulation is stronger in some than others.

4.1 Hypocholesterol Action

The lipid hypothesis purports that dietary saturated fatty acids lead to an increase in blood cholesterol levels. This may have the effect of depositing cholesterol in the arterial wall leading to atherosclerosis and possibly coronary heart disease. Some studies have hypothesised a role for the lactic microflora in systemically reducing blood lipid values.[26] However, this has not been unequivocally proven and there are contrasting data from human volunteer trials. Moreover, trials where unphysiologically high intake levels of probiotic yoghurts (> 8L/day) have not helped. Volunteer dietary trials should be carried out using a random double blind placebo procedure, with unequivocal testing of bacterial changes (eg using the genetic fingerprinting procedures mentioned earlier) and a range of human subjects.

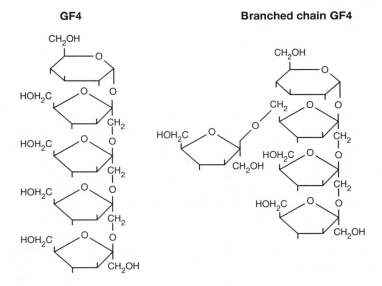

Figure 1 *Chemical structure of various fructooligosaccharides; G = glucose; F = fructose*

As the relevant blood molecules are measurable, it may be possible to ascribe mechanistic hypotheses to the role of probiotics and prebiotics in systemic reduction of blood lipids. In this case, total low density lipoprotein and high density lipoprotein cholesterol, triglycerides and apolipoproteins are relevant, with the following possible mechanisms:

- the formation of certain bacterial end products of fermentation and their subsequent absorption may affect systemic lipid and cholesterol levels
- lactic acid bacteria may be able to directly assimilate cholesterol, although this is an area of much contention
- deconjugation of bile salts
- a reduction in the absorption of cholesterol
- its conversion to other metabolites, eg coprostanol.

4.2 Gut Tumorigenesis

In humans, colorectal cancer is thought to have a bacterial origin, with the following compounds all being microbiologically derived and carcinogenic or co-carcinogenic:

- nitrosamines
- fecapentaenes
- bile acids
- heterocyclic amines
- glucuronide compounds
- various aglycones
- phenolic/indolic compounds
- nitrated polycyclic aromatic hydrocarbons
- diacylglycerol
- nitrated polycyclic aromatic hydrocarbons
- some azo compounds
- ammonia

Dietary strategies that lead to a reduced accumulation of such products may be possible. Firstly, dietary fibres and resistant starches may be fermented in the large gut to increase fecal bulk and reduce the residence time of such materials in the gut. Moreover, probiotics and prebiotics may modify the activities of enzymes that are involved in carcinogenesis, such as azoreductases, nitroreductases, β-glucuronidase, etc.[27]

Some bacterial metabolites may also offer a protective role in tumourogenesis. For example butyrate, which is a common fermentation end product, has received attention as a stimulator of apoptosis and a preferred fuel for the healthy gut mucosa. In this case, prebiotics that are fermented to increase butyrate production in the large gut may be relevant. It may also be possible to alter the metabolic traits of some microbiota components such that they do not form carcinogenic end products. The shift away from a proteolytic fermentation by clostridia and/or bacteroides to a saccharolytic one is an example.

4.3 Immunomodulation

A non-pathogenic microorganism may stimulate the immune response in such a manner that pathogenic organisms are affected. The lactic microflora are able to stimulate both non-specific and specific host defence mechanisms. Most attention in this respect has been diverted towards the intake of probiotics.[28-30]

4.4 Effects on Pathogens

The most compelling evidence for the success of probiotics and prebiotics probably lies in their ability to improve colonisation resistance, ie resistance of the effects of pathogens. Lactic acid excreting microorganisms are known for their inhibitory properties. In humans, viruses, protozoa, fungi and bacteria can all cause acute gastroenteritis. Viral infections play a major role, but bacteria are also of high significance. Cohen and Gianella[31] have summarised human bacterial pathogens into four categories. These are:

(i) Invasive organisms
ie those that occur within enterocytes or colonocytes and cause cell death such as enteroinvasive *Escherichia coli*, shigellas, campylobacters and salmonellae.

(ii) Cytotoxic bacteria
ie those which elaborate substances that directly cause cell injury, eg some shigellas, enteropathogenic *Escherichia coli* and enterohaemorrhagic *Escherichia coli*.

(iii) Toxigenic organisms
where enterotoxins adversely influence intestinal salt and water secretion, eg *Vibrio cholera*, some shigellas and enterotoxigenic *Escherichia coli*.

(iv) Adhesive bacteria
ie those which bind tightly to the colonic mucosa, such as enteroaggregative *Escherichia coli*.

The evidence for probiotic microorganisms to reduce intestinal infections, both bacterial and viral, has been reviewed.[32]

There are a number of potential mechanisms for these effects. Firstly, metabolic end products such as acids excreted by these microorganisms may lower the gut pH to levels below those at which pathogens are able effectively to compete. Also, many lactobacilli and bifidobacterial species are both able to excrete natural antibiotics which can have a broad spectrum of activity (eg lactocins, helveticins, lactacins, curvacins, nisin, bifidocin). For the bifidobacteria, our studies have indicated that some species are able to exert antimicrobial effects on various Gram positive and Gram negative intestinal pathogens.[33] This includes the Verocytotoxin strain of *Escherichia coli* 0157:H7. Other primary bacterial metabolites such as carbon dioxide, diacetyl, and hydrogen peroxide may have similar inhibitory effects. However, it should also be stated that the colonisation resistance mechanisms that the normal gut flora exerts against pathogens may also be effective against probiotics!

Other possible mechanisms of effect include competition for nutrients and other growth factors, the immunostimulation mentioned earlier, as well as competition for adhesion receptors (both lactobacilli and bifidobacteria are able to adhere to the gut epithelium possibly occupying the niche normally colonised by pathogens).

For prebiotics that increase bifidobacteria or lactobacilli towards being the numerically predominant genus in the colon, an improved colonisation resistance may result. Moreover, oligosaccharides themselves may act as anti-infective agents through the competitive occupation of pathogen colonisation/receptor sites.[34]

4.5 Lactose Intolerance

Probiotics are thought to be involved in the alleviation of symptoms caused by lactose malabsorption. In a 'lactase non-persistent' group, lactose administered in yoghurt can be utilised more efficiently than the same amount given in untreated milk. The basis for improved digestibility of lactose in yoghurt is unclear. However, possible mechanisms include a) the lactase activity of the bacteria, b) a stimulation of the host's mucosal lactase activity and c) a slower intestinal transit of yoghurt compared to milk.[35]

5 GENERAL CONCLUSIONS

The human colon is an authentic organ of nutrition and digestion that has an important role in the functional food concept. The gut microbiota contains pathogenic, benign and possible, health promoting components. Diet can be used to manipulate the gut microflora composition with probiotic, prebiotic and synbiotic (combinations of the two) products being developed to supplement, or fortify, the 'health promoting' community. Lactobacilli and bifidobacteria are the usual target organisms.

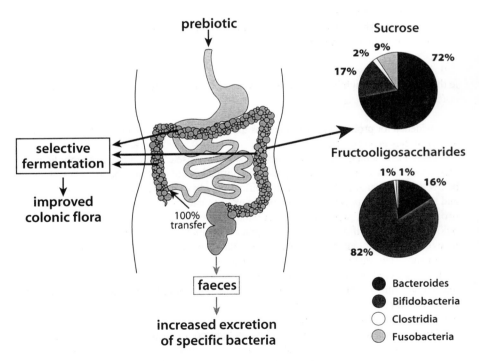

Figure 2 *The prebiotic concept and pie diagrams showing how the addition of 15g/day sucrose or fructooligosaccharides given to volunteers on a strictly controlled diet can affect the composition of fecal bacteria*

The approach of using diet to improve health is attractive to the consumer and may be a straightforward mechanism towards prophylactic aspects of gut disease. Future research should be carried out carefully and ought to take advantage of modern technologies, such that volunteer trials can be conducted that are extensive and well controlled. Recent developments in molecular techniques and efficient modelling procedures are valuable. There are three areas of interest:

* Improved knowledge of the microflora composition and the development of reliable, validated molecular-based methodologies for accurate and sensitive monitoring of species changes, eg through gene probing. This is seen as an essential step forward for determining the validity of the functional food concept in human gut microbiology.

* The fermentation of raw and processed food ingredients that are metabolised by the gastrointestinal microflora, with an assessment of the potential for flora modulation both in terms of composition and activity.

* A realistic and rational determination of the health consequences of dietary intake that targets the gut microbiota, or its individual components.

6 REFERENCES

1. G.R. Gibson and M.B. Roberfroid, *J. Nutr.*, 1995, **125**, 1401.
2. M.B. Roberfroid, *World Ingred.*, 1995, 42.
3. R. Fuller, (ed.) 'Probiotics: The Scientific Basis' Chapman & Hall, London, 1992.
4. R. Fuller, *J. Appl. Bacteriol.* 1989, **66**, 365.
5. E. Metchnikoff. 'The Prolongation of Life', William Heinemann, London, 1907.
6. H. Tissier, *Ann. Inst. Pasteur.*, 1905, **19**, 109.
7. J.H.J. Huis in't Veld and C. Shortt. 'Gut Flora and Health - Past, Present and Future', A.R. Leeds and I.R. Rowland (eds.), RSM Press, London, 1996, p. 27
8. N. Bohnhoff, B.L. Drake, and C.P. Muller, *Proc. Soc. Exp. Biol. Med.* 1954, **86**, 132.
9. F.M. Collins and P.B. Carter, *Infect. Immunol.*, 1956, **104**, 411.
10. A. Schwass, S. Sjolin, U. Trottestam and B. Aransson, *Scand. J. Infect. Dis.*, 1984, **16**, 211.
11. R.G. Kok, A. De Waal, F. Schut, G.W. Welling, G. Weenk and K.J. Hellingwerf, *Appl. Environ. Microbiol.*, 1996, **62**, 3668.
12. P. Pochart, P. Mariean, Y. Bouhnik, I. Goderel, P. Bourlioux and J.C. Rambaud, *Am. J. Clin. Nutr.*, 1992, **55**, 78.
13. B.R. Goldin, S.L. Gorbach, M. Saxelin, S. Barakat, L. Gualtiere and S. Salminen, *Dig. Dis. Sci.*, 1992, **37**, 121.
14. Y. Bouhnik, P. Pochart, P. Marteau, G. Arlet, I. Goderel and J.C. Rambaud, *Gastroenterol.*, 1992, **102**, 875.
15. H. Hidaka, T. Eida, T. Takiwaza, T, Tokunga and Y. Tashiro, *Bifid.Microflora* 1986, **5**, 37.

16. T. Mitsuoka, H. Hidaka and T. Eida, *Die Nahrung,* 1987, **31,** 427.

17. T. Mitsuoka, *Bifid. Microflora,* 1982, **1,** 3.

18. Z. Tamura, *Bifid. Microflora* 1983, **2,** 3.

19. J. Edelman and A.G. Dickerson, *Biochem. J.* 1966, **98,** 787.

20. J.J. Rumessen, S. Bode, O. Hamberg, and E.G. Hoyer, E.G, *Am.J. Clin. Nutr.* 1990, **52,** 675.

21. X. Wang and G.R. Gibson, *J. Appl. Bacteriol.* 1993, **75,** 373.

22. G.R. Gibson, E.B. Beatty, X. Wang and J.H. Cummings, *Gastroenterol.* 1995, **108,** 975.

23. R.G. Grittenden and M.J. Playne, *Trend. Food Sci. Technol.* 1996, **7,** 353.

24. I.R. Rowland and R. Tanaka, *J .Appl. Bacteriol.* 1993, **74,** 667.

25. P.C. MacGillivray, H.V.L. Finlay and T.B. Binns, *Scot .J .Med.* 1959, **4,** 182.

26. C.F. Fernandes, K.M. Shahani and M.A. Amer, *FEMS Microbiol. Rev.* 1987, **466,** 343.

27. I. Rowland, 'Gut Flora and Health - Past, Present and Future', A.R. Leeds and I.R. Rowland (eds.), RSM Press, London, 1996, p. 19.

28. E.J. Schiffrin, F. Rochat and H. Link-Amster, *J. Dairy Sci.* 1995, **78,** 1597.

29. G. Perdigon and S. Alvarez, 'Probiotics: The Scientific Basis', R. Fuller (ed.), Chapman & Hall, London, 1992; p. 146.

30. J.K. Collins, G. O'Sullivan and F. Shanahan, 'Gut Flora and Health - Past, Present and Future', A.R. Leeds and I.R. Rowland (eds.), RSM Press, London, 1996, p. 13.

31. M.B. Cohen and R.A. Gianella, 'The Large Intestine: Physiology, Pathophysiology and Disease', S.F. Phillips, J.H. Pemberton and R.G. Shorter, Raven Press Ltd. New York, 1991, p. 395.

32. G.R. Gibson, J.M. Saavedra, S. Macfarlane and G.T. Macfarlane, 'Probiotics: Therapeutic and Other Beneficial Effects', R. Fuller (ed.), Chapman & Hall, London.

33. G.R. Gibson and X. Wang, *J. Appl. Bacteriol.* 1994, **77,** 412.

34. D. Zopf and S. Roth, *Lancet* 1996, **347,** 1017.

35. P. Marteau and J.C. Rambaud, 'Gut Flora and Health - Past, Present and Future', A.R. Leeds and I.R. Rowland (eds.), RSM Press, London, 1996, p. 47.

THE EFFECT OF A NEW FERMENTED MILK PRODUCT ON PLASMA CHOLESTEROL AND APOLIPOPROTEIN B CONCENTRATIONS IN MIDDLE-AGED MEN AND WOMEN

V.A. Sessions,[1] J.A. Lovegrove,[2] T.S. Dean,[3] C.M. Williams,[2] T.A.B. Sanders,[3] I.A. Macdonald[1] and A.M. Salter[1]

[1]Department of Nutritional Biochemistry, University of Nottingham
[2]Department of Food Science, University of Reading
[3]Nutrition, Food & Health Research Centre, Kings College, University of London

1 INTRODUCTION

In recent years we have seen the emergence of the concept of functional foods. While no single definition has been adopted, most people would accept that a functional food is a dietary component which will cause specific beneficial effects on the body, making a positive contribution to the maintenance of good health and/or the prevention of disease. While this, of course, could be applied to any essential component of the diet, the definition usually implies effects over and above the amount of a nutrient required to fulfill its essential roles.

One area of particular interest is that of pre- and probiotics. These are foods which either directly contribute potentially beneficial microorganisms to the intestinal microflora (probiotics) or enhance the growth of such beneficial microorganisms (prebiotics). One health benefit which has been ascribed to a number of such foods is that of reducing risk of cardiovascular disease, by reducing plasma cholesterol. These claims have their origin in work carried out on the Massai in East Africa who have a low incidence of coronary heart disease and low plasma cholesterol concentrations despite consuming a diet high in saturated fat and cholesterol. It has been suggested that this may be a result of large intakes of fermented milk.[1] This appeared to be supported by some studies in Western subjects consuming large amounts of yogurt.[2-4] However, other human feeding studies failed to find such an effect.[5-7]

More recently it has been reported that a fermented milk product produced through the action of a bacterial culture containing a strain of *Enterococcus faecium* and two strains of *Streptococcus thermophilus* is effective in reducing plasma cholesterol at relatively modest levels of intake.[8,9] The culture was the same as that apparently isolated from the intestinal flora of elderly people living in Abkhasia in Caucasus, an area noted for the longevity of its population. In a study on Danish middle-aged men the fermented milk product reduced total plasma cholesterol by approximately 6% and low density lipoprotein (LDL) cholesterol by 10% over a period of six weeks.[8] No change was seen in high density lipoprotein (HDL) cholesterol or plasma triacylglycerol (TAG). In a follow-up study the fermented product also reduced plasma cholesterol after three months but the effect, compared to a non-fermented placebo product, was lost after six months.[9]

In the present study we investigated the effects of a milk product, fermented with the same bacterial culture described above, on plasma cholesterol and apolipoprotein (apo) B concentrations in men and women with mildly elevated plasma cholesterol concentrations living in England. The study was multicentred, placebo controlled and double blind.

2 MATERIALS AND METHODS

2.1 Subjects

173 men and women, from three different English centres (Nottingham, Reading and London), between the ages of 30 and 55 years, were selected for the study on the basis of having blood cholesterol concentrations of between 5.2 and 7.8mM. Subjects were free of any history of cardiovascular disease or of metabolic disorders including, diabetes or thyroid disorder. They were not receiving any lipid-lowering medication, were normotensive and non-obese (BMI<30.0 kg/m^2).

2.2 The Test and Placebo Products

The test product was a bacterially fermented milk product which is fermented with a blend of *Enterococcus faecium* and two strains of *Streptococcus thermophilus*. The study was double blind and placebo controlled with a chemically produced milk product of identical composition but which was produced with an organic acid (delta-gluco-lactone). Both the test and the placebo products were flavoured with strawberry. The nutritional content of both products was: 4.5g protein, 6.0g carbohydrate and 1.3g fat per 100g with an energy content of 23KJ/100g. Each volunteer was required to consume 200ml (provided in one pot) of either the test or the placebo product daily for 12 weeks. The allocation of subjects to the test or the placebo group was random.

2.3 The Trial

During the 12 week study, volunteers were required to give fasting samples of blood for determination of plasma cholesterol and apolipoprotein B. Two baseline samples (7 days apart) were taken and then further samples at 6 weeks and 12 weeks. The mean of the two baselines values was used as the initial value.

Plasma cholesterol was determined enzymatically on a COBAS MIRA autoanalyser using the Olympus system reagent 5000 kit and apoB was determined by immunoturbidometry using a kit from Sigma-Aldrich Co. (Dorset, UK).

3 RESULTS

Of the 173 volunteers who started, 160 completed the 12-week study (82 females and 78 males). On a self-reporting basis compliance was good, with only two subjects reporting missing seven or more yoghurts throughout the duration of the study. Some minor gastrointestinal complaints including indigestion, looser stools, diarrhoea, constipation and flatulence were reported though these tended to be greater in the placebo than the test group. None of these were deemed serious enough to withdraw any subjects from the study.

Figure 1 shows the plasma cholesterol and apo B concentrations in the two groups with time. It can clearly be seen they were initially well matched with respect to both parameters. No significant change was seen in either cholesterol or apoB with time and no significant differences were seen between the test and placebo groups at any time point. When the data were analysed separately for men and women no significant effects were seen (data not shown). Data were also re-analysed excluding subjects who had taken antibiotics during the trial (n=13), and no effect was seen (data not shown).

4 DISCUSSION

Many of the health claims associated with functional foods are currently regarded with considerable suspicion by the consumer. Such foods are often retailed without the rigorous scientific testing of efficacy that would be required if the product were marketed as a pharmaceutical agent.

The fermented milk product used in the current study is, however, an exception to this generalization. Two well performed clinical trials have already been published in the scientific literature. The first, performed on middle-aged Danish men, clearly indicated a cholesterol-lowering effect and further showed that this was restricted to the potentially atherogenic LDL fraction.[8] These men had comparable starting plasma cholesterol concentrations to the subjects in the present study. They were a more homogenous population in terms of age, all being born in the same year. The study was of similar design but over a shorter duration of only six weeks. It should, however, be noted that in the present study we failed to see an effect even after this time period.

The results of the second study[6] were a little less clear cut. The study included 90 subjects of both sexes between the ages of 50 and 70 years. Initial plasma cholesterol concentrations were again similar to those of the subjects in the present study. Significant differences between the test and placebo groups were seen up to a period of three months but by six months no such difference was apparent. This was due, at least in part, to the fact that cholesterol concentrations had also fallen in the placebo group. Interpretation of these results was also made difficult by the fact that during the second half of the study the concentration of *Enterococcus faecium* in the test product dropped from 10^8/ml to 10^4-10^5/ml. As *Streptococcus thermophilus* does not survive to any degree in the small intestine it appears likely that *Enterococcus faecium* is responsible for the cholesterol-lowering effects. Thus the drop in concentration of this bacteria may have been responsible for a reduction in the efficacy of the fermented milk product. Neither of these confounding effects were seen in the present study. Thus, no changes were seen in either the placebo or test group and the bacterial content of the fermented milk remained constant.

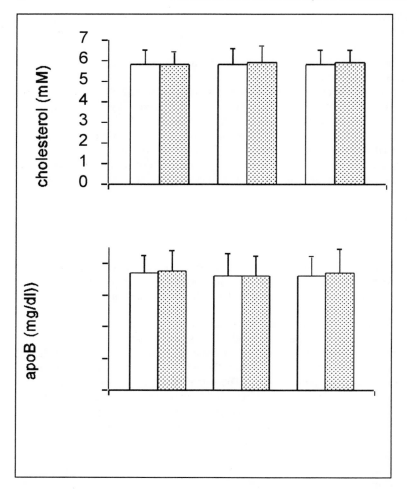

Figure 1   *Plasma cholesterol and apolipoprotein (apo) B concentrations in subjects consuming either the bacterially fermented (white bars) or chemically produced (shaded bars) milk product. Results are presented as mean ± standard deviation.*

Why did we not see a cholesterol-lowering effect of the fermented product in our study? It could be speculated that this is a result of inherent differences between the populations studied. Consumption of yogurt in England is considerably lower than in Denmark, particularly amongst men. Thus, while Danish subjects were likely to be replacing yogurt which they were habitually consuming, the 200ml of the fermented (or chemically produced) milk product is likely to have had a more significant impact on the overall diet of the English subjects. It is also possible that there are inherent differences in the intestinal flora of the populations of the two countries which may have an impact on the efficacy of the fermented product.

The results of this study, combined with those of the other two reported in the literature, highlight the problems associated with verifying the health claims of functional foods. Such foods are generally aimed at a diverse, often global, market. It is quite likely that their effects will vary depending on age, sex, ethnicity and habitual dietary intakes. However, without a battery of rigorously performed, independent studies such as those outlined here, the consumer has little protection from unsubstantiated health claims being used to promote specific products.

5 REFERENCES

1. G.V. Mann and A. Spoerry, *Am. J. Clin. Nutr.*, 1974, **27**, 464.
2. G.V. Mann, *Atherosclerosis*, 1977, **26**, 335.
3. G. Hepner, R. Fried , S. St Jeor, L. Fusetti and R. Morin, *Am. J. Clin. Nutr.*, 1979, **32**, 19.
4. T.L. Bazzarre, S. Liu Wu and J.A. Yuhas , *Nutr. Rep. Internat.*, 1983, **28**, 1225.
5. J.E. Rossouw, E-M. Burger, P. Van der Vyver and J.J. Ferreira, *Am J. Clin. Nutr.* 1981, **34**, 351.
6. L.U. Thompson , D.J.A. Jenkins, M.A.V. Amer , R. Reichert, A. Jenkins and J. Kamulsky, *Am J. Clin. Nutr.* 1982, **36**, 1106.
7. D.J. McNamara, A.E. Lowell and J.E. Sabb, *Atherosclerosis*, 1989, **79**, 167.
8. M. Agerbaek, L. U. Gerdes and B. Richelsen, *Eur. J Clin Nutr.*, 1995, **49**, 346.
9. B. Richelsen, K. Kristensen and S.B. Pedersen, *Eur. J Clin Nutr.*,1996, **50**, 811.

BENEFITS AND RISKS OF *ENTEROCOCCUS FAECIUM* AS A PROBIOTIC

J.M.T. Hamilton-Miller and S. Shah

Department of Medical Microbiology
Royal Free Hospital School of Medicine
London NW3 2QG

1 INTRODUCTION

1.1 Definitions

There have been several definitions of the word probiotic since it was first coined (reputedly in 1965[1]) with the meaning of an inanimate substance that stimulates microbial growth. Parker[2] appears to have been the first to use the word to describe live microbes taken as a food supplement, and the definition has now moved on to include preparations containing microbial cell components in the absence of viable organisms. The continuing debate on definitions[3] suggests that none has received universal acclaim. The following discussion will be confined to preparations that contain live bacteria, the purpose of which, ostensibly, is to redress the microbial balance of the intestinal flora, to the ultimate advantage of the host.

Probiotic bacteria have traditionally been selected from among the lactic acid organisms by the application of strict criteria;[4] they include lactobacilli, bifidobacteria and streptococci. Due to their supposed safety and potential beneficial effects, probiotic strains are often termed 'friendly bacteria', to distinguish them from other groups in the gut flora that have greater potential to be harmful (eg members of the Enterobacteriaceae).

1.2 Background to Study

Probiotic bacteria are available to the general public as supplements, which consist of freeze-dried preparations of bacteria (in powder form, in capsules or as tablets), and in functional foods (eg yoghurts), where vegetative bacteria are presented in their spent culture medium.

The literature shows that yoghurts may be an unreliable source of friendly bacteria,[5] and that supplements can contain not only lower numbers of viable organisms than claimed, but may have species other than those listed, and may lack those that are listed.[6-9] There appear to be no generally acceptable standard procedures for Quality Assurance for probiotic supplements, as well as a shortfall in the usual customer protection measures applicable to foodstuffs.

The authors' particular concern, as Medical Microbiologists, has been the finding of viable *Enterococcus faecium* in several probiotic supplements and functional foods. Concern is heightened when this species is present but not announced on the product's label, or when its obsolete name (see below) is used.

2 PRESENCE OF *E.FAECIUM* IN PROBIOTICS

2.1 Authors' Findings

2.1.1. Methods and Materials. We have examined a total of 47 different products (36 probiotic supplements, nine 'live' or 'bio' yoghurts and two fermented milks) for the presence of enterococci. Most of these (35) were obtained in the UK, from retail pharmacies, health food shops or by mail order. The remainder (12) were from continental Europe (eight from France, two from Denmark, two from Germany), kindly supplied by colleagues. Depending on the formulation of the individual product, 1g powder, 1ml fluid, 1 crushed pre-weighed tablet or the weighed contents of a capsule was suspended in 10ml saline, mixed with a vortex mixer and allowed to settle for 20 minutes on the bench. The supernatant fraction (0.1ml) and decimal dilutions thereof was spread onto m-Enterococcus agar (Difco), incubated overnight at 37 ^{0} C and colonies counted. Enterococci were readily distinguishable from other species present, growing as small colonies with a magenta centre. Identification was by API Strep kits, or by courtesy of the Streptococcal Reference Laboratory, Colindale. Viable counts were expressed as colony forming units (cfu)/g.

2.1.2 Results. Enterococci were found in 13 of the products tested (28%). They were more likely to be found in probiotic supplements (11/36 = 30.6%) than in functional foods (2/11 = 18.2%), but this difference is not statistically significant (chi squared). Numbers ranged from 2.5×10^9 - <200/g (in one product enterococci were isolated only after enrichment), with a median of 1.4×10^4/g. Examination of these figures showed two distinct populations—the six products (all probiotic supplements) yielding the largest numbers of enterococci (range 2.5×10^9- 1×10^7/g) were all labelled as containing these organisms (usually, however, using the obsolete nomenclature), while the other seven (two bioyoghurts and five supplements) had much smaller numbers (range 1.4×10^4 - < 200/g) and in no case was the fact that enterococci were present indicated on the label.

The 13 strains of enterococci isolated were identified as *E.faecium* (12) and *E.faecalis* (1). All were sensitive to ampicillin, chloramphenicol, imipenem, vancomycin, teicoplanin, trimethoprim, ciprofloxacin and tetracycline, and all resistant to cephalexin and fosfomycin. There were varying amounts of resistance to other clinically useful antibiotics—erythromycin (54%), gentamicin (69%), cefotaxime (77%) and rifampicin (81%).

2.2 Literature Reports

E.faecium has been reported to be present in probiotics available in the USA. Hughes & Hillier[8] found between 10 and 10^5 cfu/g in 10 of 11 probiotics tested, this information being 'on label' for only one product. More recently, Coque *et al.*[10] isolated this species (numbers not specified) from five products, and Alcid *et al.*[11] from two, both 'on label' (10^6 and 10^9 cfu/g). Of great importance, one of the strains from the latter study was found to contain the *valiB* gene, and was therefore resistant to vancomycin (see below). *E.faecium* SF68 is available in Italy.[12-16]

3 POTENTIAL BENEFICIAL EFFECTS OF *E.FAECIUM*

Supposedly probiotic effects of live *E.faecium* in man have been reported, using one strain, SF68. Four of these in English [12-15] involved the treatment of enteritis, almost always of unspecified aetiology, and some of the patient groups were far from typical (eg mainly alcoholic suffers of tuberculosis,[14] children admitted to hospital[12]). Another, more convincing, study was in the treatment of hepatic encephalopathy in cirrhotics.[16] The positive results found in these five studies were not very great compared to placebo or more conventional treatment, and were attributed to colonization of the gut by the probiotic organism.   However, no bacteriological studies were done to confirm this. Other studies have been reported with this strain in different languages (see[15]).

Even if the evidence of SF68 as a useful probiotic organism is accepted at face value, the properties of this one strain cannot be extrapolated to all strains of *E.faecium*, nor results in one patient group to healthy individuals.

It has been claimed that the strain of *E.faecium* (K77-D) contained in the ('Causido') culture might reduce serum cholesterol levels,[17] but the evidence did not satisfy all the UK authorities.   The product containing this strain has been withdrawn from the UK market, although it is available elsewhere.

A further study[18] involved the use of viable *E.faecalis*, given for a six month period to sufferers of chronic, relapsing tonsillitis, with a beneficial effect. The patients' improvement was explained in terms of immunological modulation. It would be interesting to know whether the outcome would have been the same had killed cells been used, as *E.faecalis* has been reported[19] to be an immunostimulant in mice.

Thus, the evidence for a probiotic role in man for enterococci, and *E.faecium* in particular, is not very strong, even for the defined strain SF68. The latter is said to be resistant to many antibiotics,[15] and so clearly does not correspond to any of the strains we isolated. We submit that there is little if anything to suggest that this species has a role to play in maintaining good health in already healthy people.

4 POTENTIAL DANGERS OF *E.FAECIUM*

The fact that *E.faecium* used to be known as *Streptococcus faecium*[20] (and still is, according to labels on many probiotics) may have created a false sense of security concerning the potential pathogenicity of this species. Unlike other species either now and formerly belonging to the genus Streptococcus—such as *Lactococcus lactis* (originally *S.lactis*, Lancefield group N) and *S.salivarius* var *thermophilus*) - enterococci in general and *E.faecium* in particular now present a real hazard in hospitals.

Enterococci have had a rapid rise to infamy over the past two decades. They are now the second most frequent cause of nosocomial urinary infections and the third commonest reason for surgical wound infections and bacteraemia in the USA.[21]

These organisms, and especially *E.faecium*, have proved extremely adept at acquiring antibiotic resistance, most seriously to vancomycin.[22] The situation has now arisen when we are seeing infections with untreatable organisms. Mortality arising from enterococcal infections even with sensitive strains is in excess of 30%, and patients surviving have to spend on average an extra 39 days in hospital as a result.[23]

Immunocompromized patients (eg transplant recipients, and those undergoing chemotherapy for a malignancy) are at greatest risk. Patients are almost always infected from their own endogenous flora, colonization preceding infection. A pool of organisms in the gut acquires resistance and is then selected for by the use of broad-spectrum antibiotics. This process may be short-circuited by breaking the chain. Thus, preventing the dissemination of *E.faecium* would be a useful first step.

The strains isolated here are harmless *per se*, and indeed the Causido strain K77- D was cleared by MAFF. Enterococci isolated here were sensitive to many antibiotics, but at least two probiotic strains of *E.faecium*, sold in U5A[11] and Italy (5F68)[15], are known to be resistant. No consideration appears to have been given by MAFF or others to the possibility of resistance being acquired by K77-D or any other enterococcus once the bacteria have entered the large intestine and encountered many other organisms. This question could readily be answered by simple experiments, but these have not only not been done but apparently are not being contemplated.

5 CONCLUSIONS

On one hand, enterococci are without doubt potentially dangerous pathogens. On the other hand, some would claim that *E.faecium* has useful probiotic properties. The evidence for the latter claim does not convince that its use makes a significant contribution to health. The wide dissemination of enterococci through the population serves to increase the intestinal pool and thereby gives greater opportunities for resistance acquisition and the initiation of infection in hospitals.

At the very least, labelling should be clarified so that consumers are fully aware which products contain enterococci. Discussions with wholesalers of probiotics have disclosed much ignorance about the potential dangers of enterococci, and lack of awareness of the microbiological content of finished products as opposed to starter cultures. In half the preparations from which we isolated enterococci we suspect that these were present due to hitherto unsuspected contamination, which must raise questions about Quality Assurance procedures.

6 REFERENCES

1. D.M.Lilly and R.H.Stillman, *Science*, 1964, **147**, 747
2. R.B.Parker, *Anim. Nutr. Hlth*, 1974, **29**, 4.
3. R.Fuller, P.J.Heidt, V.Rusch and D. van der Waaij (eds),"Probiotics: Prospects of Use in Opportunistic Infections", Institute for Microbiology and Immunology, Herborn-Dill, 1995.
4. Y-K Lee and S.Salminen, *Trends Fd Sci.Technol*, 1995, **6**, 241.

5. D. Miller, *Which?*, April 1993, 38.
6. S.E.Gilliland and M.L.Speck, *J. Fd Protectn*, 1977, **40**, 760.
7. S.E.Gilliland, *Oklahoma Agr. Exp. Station Misc. Pub*, 1981, **108**, 61.
8. V.L.Hughes and S.L.Hillier, *Obstet. Gynecol*, 1990, **75**, 244.
9. J.M.T.Hamilton-Miller, S.Shah and C.Smith, *Brit. Med. J.*, 1996, 312, 55.
10. T.M.Coque, J.F.Tomayko, S.C.Ricke, P.C.Okhyusen and B.E.Murray, *Antimicr. Ag. Chemother*, 1996, **40**, 2605.
11. D.V.Alcid, M.Troke, S.Andszewski and J.F.John, *Abs. Meet. Inf. Dis. Soc. Amer.*, 1994, 123.
12. G.Bellomo, A.Mangiagle, L.Nicastro and G.Frigerio, *Curr Ther. Res*, 1980, **28**, 927.
13. E.Camarri, A.Belvisi, G.Guidoni, G.Marini and G.Frigerio, *Chemotherapy*, 1981, **27**, 466.
14. M.Borgia, N.Sepe, V.Brancato and R.Borgia, *Curr. Ther. Res*, 1982, **31** ,265.
15. P.F.Wunderlich, L.Braun, I.Fumagalli, V.D'Apuzzo, F.Heim, M.Karly, R.Lodi, G.Pollita, F.Vonbank and L.Zeltner, *J. Int. Med. Res*, 1989, **17**, 333.
16. C.Loguerico, R.Abbiati, M.Rinaldi, A.Romano, C.D.V.Blanco and M.Coltorri, *J.Hepatol*, 1995, **23**, 39.
17. M.Agerbaeck, U.Gerdes, and B.Richelsen, *Eur. J. Clin. Nutr*, 1995, **49**, 346.
18. S.Kalinski, *Fortsch. Med*, 1986, **43**, 843.
19. K.Satonaka, K.Ohashi, T.Nohmi, T.Yamamoto, S.Abe, K.Uchida and H.Yamaguchi, *Microbiol. Immunol*, 1996, **40**, 217.
20. K.H.Schleifer and R.Kilpper-Balz, *Int. J. Syst. Bact.* 1984, **34**, 31.
21. T.G.Emori and R.P.Gaynes, *Clin. Microbiol. Rev.*, 1993, **6**, 428.
22. M.B.Edmond, J.F.Ober, J.D.Dawson, D.L.Weinbaum and R.P.Wenzel, *Clin. Inf. Dis.*, 1996, **23**, 1234.
23. S.L.Landry, D.L.Kaiser and R.P.Wenzel, *Am. J. Inf. Contr.*, 1989, **17**, 323.

CLINICAL EFFICACY OF A HUMAN *LACTOBACILLUS* STRAIN AS A PROBIOTIC

M. Saxelin,[1] S. Salminen[2] and E. Isolauri[2]

[1]Valio R&D, Helsinki
[2]University of Turku, Finland

1 INTRODUCTION

Lactic acid bacteria form a source of functional ingredients for products designed for the nutritional treatment of intestinal dysfunction. It is generally accepted that yoghurt alleviates lactose intolerance and is effective in the treatment of certain intestinal disorders. Since there is a clear difference in the characteristics of different strains within a bacterial species, we have concentrated on using a defined probiotic strain, *Lactobacillus rhamnosus* strain GG (ATCC 53103), in a number of clinical studies. In this presentation a short review is given of the clinical studies conducted with *Lactobacillus* GG. The main functional and health effects of *Lactobacillus* GG are that it:

- colonises the human digestive tract and balances the intestinal microflora
- prevents and treats intestinal disorders
- enhances natural intestinal resistance
- enhances recovery in cases of milk allergy.

2 INTESTINAL COLONISATION AND STABILISATION OF THE MICROFLORA

Lactobacillus GG tolerates acid conditions in the stomach and the bile acids in the small intestine better than traditional yoghurt bacteria. It is able to survive passage through the gastrointestinal tract and can be recovered in viable form from fecal samples. It has been possible to determine the colonising daily doses of the strain both in various types of dairy product and in freeze-dried powder form.[1] The attachment of *Lactobacillus* GG on human buccal cells and on cultured intestinal cells has been shown *in vitro*.[2] The strain was also recovered in biopsy samples of the colon after oral consumption of a LGG-product for two weeks and subsequent intestinal colonoscopy.[3]

2.1. Intestinal Microecology

In healthy subjects a high consumption of *Lactobacillus* GG has been shown to increase the 'beneficial' bacterial flora (lactobacilli and bifidobacteria) and to reduce certain groups of clostridia.[4] A balancing effect on the disturbed intestinal microflora was seen in a study with children with shigellosis. The children were randomised into three groups, and treated with (i) trimethoprim-sulfamethoxazole (TMP), (ii) TMP+LGG and (iii) LGG. In the group treated with TMP only, the levels of fecal anaerobic bacteria and lactobacilli were low. In the groups treated with LGG and TMP+LGG, the levels of lactobacilli were normal and the levels of anaerobic bacteria were also normalised.[5]

The balancing effect on the intestinal microecology is reflected in the activity of the fecal enzymes. Changes in intestinal enzyme activities indicate changes in the numbers of intestinal bacteria. High urease activity was detected during rotavirus diarrhoea and in juvenile chronic arthritis. *Lactobacillus* GG treatment reduced urease activity in these intestinal disorders.[6,7] In viral disorders a secondary bacterial infection may follow the osmotic diarrhoea caused by rotavirus, and this may be inhibited by the balancing effect of lactobacilli. The mechanism of the inhibition of urease-producing bacteria may be related to a specific antimicrobial effect of *Lactobacillus* GG[6] and its influence on the acidity of the intestinal content.

2.2. Fecal Hydrolytic Enzymes

High activity of hydrolytic enzymes in the colon is thought to increase the risk of colon cancer.[9] During human dietary consumption of products with *Lactobacillus* GG the activity of such enzymes (β-glucuronidase, nitroreductase, glycocholic acid hydrolase) decreased significantly.[10,11] In a placebo-controlled animal study, rats fed with a high-corn-oil diet and injected weekly with a carcinogenic compound (DMH) produced less intestinal tumours when their diet contained *Lactobacillus* GG.[12]

3 TREATMENT OF INTESTINAL DISORDERS

3.1. Acute Gastroenteritis in Childhood

Rotavirus is the most important infecting agent in acute gastroenteritis in infants and young children. The accepted treatment of acute diarrhoea is based on rehydration, preferably oral, and new supporting treatments would be appreciated. The probiotic therapy of acute gastroenteritis with *Lactobacillus* GG (after oral rehydration) has been studied in five placebo-controlled studies by Isolauri and coworkers.[13,14] In all five of these studies the mean duration of diarrhoea in hospital was from 1.1 to 1.8 days in the *Lactobacillus* GG groups, vs. 2.4 to 2.6 days in the placebo groups (p-values <0.001). *Lactobacillus* GG was as effective in a fermented milk product as in freeze-dried powder form, and in contrast to another *Lactobacillus* strain and to yoghurt starter strains, it was the only one to shorten the duration of the diarrhoea. Heat-inactivated bacteria also shortened the duration of diarrhoea, but did not have a similar immune response to live bacteria.

In addition to well-nourished children, faster recovery from diarrhoea was recorded in hospitalised children in Thailand[7] and Pakistan[8] (p-values <0.055 and <0.01 respectively). In these placebo-controlled studies, *Lactobacillus* GG was shown to be effective in the treatment of watery diarrhoea but not in cases with bloody diarrhoea.

3.2. Other Intestinal Disorders

During antibiotic therapy the intestinal microflora is generally disturbed and intestinal symptoms are common. The effect of *Lactobacillus* GG in the prevention of side-effects of erythromycin therapy was studied with healthy volunteers in a placebo-controlled study. During one-week therapy the incidence of diarrhoea was lower in the *Lactobacillus* GG group (p<0.05), and other intestinal disorders (stomach pain, abdominal distress and nausea) were also less frequent in the LGG group. Fecal counts of *Lactobacillus* GG were at a detectable level in 75% of the volunteers.[17] Pseudomembraneous colitis caused by *Clostridium difficile* is a chronic and often relapsing diarrhoea which may follow antibiotic treatment. The frequency of recurrent diarrhoea after its treatment with metronidazole or vancomycin is approximately 20%, and other methods of treatment are based on microbial therapy.[18] The successful treatment of recurrent *C. difficile* colitis with *Lactobacillus* GG has been reported in 37 cases.[18,19]

Lactobacilli are popular in the prevention of traveller's diarrhoea, although their effect is not well documented. The prevention of traveller's diarrhoea with *Lactobacillus* GG was evaluated in two placebo-controlled studies. In a study with North-Americans travelling to several destinations in developing countries, the risk of getting diarrhoea was significantly reduced with *Lactobacillus* GG (protection rate 47%, p=0.05).[20] In another study with Finnish tourists travelling to Turkey, the prevention of diarrhoea with *Lactobacillus* GG was significant in one destination (protection rate for one-week period 39.5%, p<0.05) but not in the other.[19] The reasons for the discrepancy of the results may be too low a dose of the probiotic bacteria and different loads of infective bacteria.

4 INTESTINAL EFFECTS

4.1. Enhancement of Natural Resistance

Lactobacilli enhance the colonisation resistance of the intestine against infection by balancing the microbial flora. Another important factor is the immune response evoked by bacteria in the gut-associated lymphoid tissue. A prominent immunoglobulin-secreting cell (ISC) response was measured during rotavirus diarrhoea, especially in IgM class. *Lactobacillus* GG was shown to double the production of ISC. By the follow-up this difference had vanished. The number of IgA specific antibody-secreting cells (sASC) was low in the acute phase of diarrhoea but at the three-week follow-up the specific rotavirus response in IgA sASC was significantly more common in patients in the *Lactobacillus* group (90%) than in the placebo group (46%, p=0.006).[13] This specific rotavirus response was not seen with another *Lactobacillus rhamnosus* strain (Lactophilus).[14] Feeding with certain, but not all, *Lactobacillus* strains thus seems to improve immunity against reinfections.

4.2 Stabilisation of the Mucosal Barrier

The absorption of macromolecules across the gut mucosa is increased during intestinal disorders such as acute diarrhoea. In experimental studies *Lactobacillus* GG influenced the transport of food antigens through the gut mucosa, increasing the transport via Payer's patches. This led to an enhanced local IgA response and stabilisation of the integrity of the jejunal mucosa, diminishing the transport of intact molecules through the mucosa.[21,22] There is also evidence that *Lactobacillus* GG positively influences intestinal permeability during chronic inflammation. Such studies have been conducted in patients with Crohn's disease, resulting in an increased local IgA response to food antigens.[23]

5 NOVEL TREATMENT OF MILK ALLERGY

During a variable period after birth, the mucosal immune system is functionally immature. Consequently, the immune exclusion functions are incompletely developed in early infancy, manifesting intransigently increased intestinal permeability. The binding of antigens to the immature gut microvillus membrane is increased compared to the mature mucosa, which has been shown to correlate with the increased uptake of intact macromolecules. An increased antigen load on the immature immune system may evoke aberrant antigen transfer and immune responses and lead to sensitization. Thus early antigen exposure may result in priming for immune responses instead of inducing tolerance.

Development of the immunophysiological regulatory mechanisms in the gut mucosa depends on the establishment of a normal intestinal microflora and the introduction of dietary antigens. The demonstration that the gut microflora is an important constituent in the intestine's mucosal barrier has introduced a novel therapeutic intervention.[24] As was shown in the experimental studies, the mucosal barrier was normalised with *Lactobacillus* GG. The beneficial effect of the strain on infants with cows' milk allergy was shown in a placebo-controlled study, where the infants consumed a milk-free diet of extensively hydrolysed whey formula, either supplemented (n=13) or not (n=14) with *Lactobacillus* GG. After one month of treatment the children had less severe atopic eczema in the *Lactobacillus* GG group compared to the placebo group (p=0.05). After two months the recovery was equal in both groups. The concentrations of fecal α-1 antitrypsin and TNF-α decreased significantly in the *Lactobacillus* group, but not in the placebo group, during one month's management, indicating recovery from the intestinal inflammation.[25] Hydrolysis of milk caseins with *Lactobacillus* GG proteases downregulated the production of IL-4 cytokine *in vitro*[26] and suppressed the induced lymphocyte proliferation,[27] indicating that the mechanism behind the beneficial effect was partly in the regulation of cytokine production.

6 CONCLUSION

The balancing effect on the intestinal microflora of probiotic bacteria, the enhancement of the natural intestinal defence, the beneficial effects on intestinal disorders, and the prevention and treatment of intestinal infections are potentially realistic today. However, it should be remembered, that not all strains in a bacterial species are equal, and the health effects should be evaluated with each strain separately. New applications for probiotics, such as the alleviation of food allergy, can also be found, but basic research should be continued alongside applied studies.

7 REFERENCES

1. M. Saxelin, *Nutrition Today*, 1996, **31**, Supplement 1, 5S.
2. S. Elo, M. Saxelin and S. Salminen, *Lett. Appl. Microbiol.*, 1991, **13**, 154.
3. M. Alander, R. Korpela, M. Saxelin, T. Vilpponen-Salmela, T. Mattila-Sandholm and A. von Wright. *Lett. Appl. Microb.* 1997, in press.
4. Y. Benno, F. He, M. Hosoda, H. Hideo, T. Kojima, K. Yamazaki, H. Iino, H. Mykkänen and S. Salminen, *Nutrition Today*, 1996, **31**, Supplement 1, 9S.
5. E. Sepp, E. Tamm, S. Torm, I. Lutsar, M. Mikelsaar and S. Salminen, *Microecol and Therapy*, 1995, **23**, 74.
6. M. Silva, N.V. Jacobus, C. Deneke and S.L. Gorbach, *Antimicrob. Agents Chemother.* 1987, **31**, 1231.
7. A.R. Pant, S.M. Graham, S.J. Allen, S. Harikul, A. Sabchareon, L. Cuevas and C.A. Hart, *J Tropic. Pediatrics*, 1996, **42**, 162.
8. S. Raza, S.M. Graham, S.J. Allen, S. Sultana, L. Cuevas and C.A. Hart, *Pediatr. Infect. Dis. J.*, 1995, **14**, 107.

THE INFLUENCE OF *BIFIDOBACTERIUM LONGUM* AND INULIN (RAFTILINE HP) ON COLONIC NEOPLASTIC LESIONS AND GUT BACTERIAL METABOLISM: PROBIOTIC, PREBIOTIC AND SYNBIOTIC EFFECTS

C. Rumney, J. Coutts, L. Lievense[1] and I. Rowland

BIBRA International, Carshalton, Surrey, SM5 4DS
[1]Unilever Research Vlaardingen, Olivier van Noortlaan 120, 3133 AT Vlaardingen, The Netherlands.

1 INTRODUCTION

Lactic acid producing bacteria (probiotics) and substrates that preferentially stimulate the indigenous population of these organisms in the gut (prebiotics) are claimed to have potential health benefits. We have been studying the potential cancer-preventing properties of probiotics and prebiotics using short term assays for colon carcinogenesis.

We have investigated the effect of *Bifidobacterium longum* (a probiotic) and a derivative of inulin (Raftiline HP; a purported prebiotic) on the formation of aberrant crypt foci (ACF) in the colon of rats. ACF are induced by treatment of rats with colon carcinogens and are considered to be early, pre-cancerous lesions.[1] Cecal pH and various bacterial enzyme activities and metabolites thought to play a role in colon carcinogenesis were also determined.

2 MATERIALS AND METHODS

Raftiline HP was was purchased from Orafti, Belgium. The model colon carcinogen azoxymethane (AOM) was purchased from Sigma Chemical Company, Poole, Dorset, UK.

Bifidobacterium longum 25, provided by Unilever Research Laboratory, Vlaardingen, was supplied as a freeze-dried powder containing approximately 4×10^{10} colony forming units per gram.

Four groups of 15 male Sprague Dawley rats, fed a basal high fat (25%w/w) diet (CO25) were given 2 s.c. doses of AOM (12.5 mg/kg) one week apart. Two weeks after the second dose, the rats were transferred to the experimental diets:

1. CO25
2. CO25 + *B. longum* (7×10^8 cells/g diet)
3. CO25 + Inulin (5% w/w)
4. CO25 + *B. longum* + Inulin

The rats were killed by cervical dislocation 12 weeks later and the cecal pH measured *in situ* prior to the contents being removed for subsequent analysis of β-glucuronidase activity[2] and ammonia concentration.[3] The colons were excised and ACF enumerated after staining with methylene blue.[4]

Results were subjected to analysis of variance using the Minitab statistical package. Individual means were compared using the Least Significant Difference criterion.[5]

3 RESULTS

Rats fed the diets containing either Raftline alone or with *B. longum* exhibited significantly decreased (P<0.001) cecal pH. The cecal concentration of ammonia was significantly decreased (P<0.01) by feeding either Raftiline or *B.longum* alone, but an even greater reduction (P<0.001, by comparison with the control group) was observed in rats given the combination of Raftiline and *B.longum*. β-Glucuronidase was also significantly decreased in all groups on experimental diets, although the greatest effect (P<0.01) was seen in the group fed Raftiline and *B. longum* together.

B. longum incorporated into the diet alone caused a significant (P<0.05) decrease in the number of small (those with 1-3 crypts per focus), AOM-induced ACF, whilst Raftiline elicited an even greater reduction (P<0.01). The largest effect however, was seen in those rats fed the Raftiline and *B. longum* together, with an approximate 80% reduction in small ACF (P<0.001).

No effect was seen in rats given Raftiline or *B. longum* alone on large ACF (those with four or more crypts per focus), whereas the combination of the two in the same diet produced a significant reduction (P<0.05) of approximately 60%.

4 CONCLUSIONS

Dietary Raftiline decreased cecal pH, which is consistent with consumption of a highly fermentable carbohydrate. *B. longum* feeding however, had no effect on cecal pH. Consumption by rats of either *B. longum* or Raftiline decreased the activity of β-glucuronidase and ammonia concentration in cecal contents, both of which have been associated with colon carcinogenesis in experimental animals. Furthermore, combined treatment with Raftiline and *B. longum* was more effective than separate administration, suggesting a synbiotic effect.

B. longum or Raftiline alone in the diet reduced numbers of AOM-induced, small ACF in the colon, but the effect was enhanced by the combined treatment of probiotic and prebiotic together. The significance of ACF numbers has been the subject of much debate, however some workers suggest that the number of large ACF (those with four or more aberrant crypts per focus) is more indicative of eventual tumour incidence. In the present study, only the combined treatment decreased numbers of large ACF.

To conclude, Raftiline appears to act synergistically with *B. longum* (presumably by providing a carbon/energy source for the latter) to reduce a number of biomarkers of colon cancer risk.

5 ACKNOWLEDGEMENT

The authors wish to acknowledge the financial support for this research programme by EU (AIR1-CT92-0256) and MAFF.

6 REFERENCES

1. T.P. Pretlow, B.J. Barrow, W.S. Ashton, M.A. O'Riordan, T.G. Pretlow, J.A. Jurcisek and T.A. Stellato, *Cancer Res.*, 1991, **51**, 1564-1567.
2. A.K. Mallett, A. Wise and I.R. Rowland, *Fd. Chem. Toxicol.*, 1984, **22**, 415-418.
3. A. Wise, A.K. Mallett and I.R. Rowland, *Nutr. Cancer*, 1983, **4**, 267-272.
4. R.P. Bird, *Cancer Lett.*, 1987, **37**, 147-151.
5. G.W. Snedecor and W.G. Cochran, 'Statistical Methods 6th Edition', Ames: Iowa State University Press.

EFFECT OF OLIGOFRUCTOSE FEEDING ON GLUCOSE AND INSULIN LEVELS IN THE RAT

N. Kok, M. Roberfroid and N. Delzenne

Unité de Biochimie Toxicologique et Cancérologique
Département des sciences pharmaceutiques
Université Catholique de Louvain
UCL-BCTC 7369
73 Avenue Mounier
B- 1200 Brussels, Belgium

1 INTRODUCTION

Today, consumers look more and more for foods with high nutritional value, safe in terms of toxicity but also foods which are beneficial for health. Therefore, the concept of functional foods has emerged. In this regard, fructooligosaccharides and in particular oligofructose have interesting nutritional properties that make these compounds candidates for classification as health-enhancing functional food ingredients.

Oligofructose (OFS) is a natural common food ingredient but can also be prepared by enzymatic hydrolysis of chicory inulin. This commercially available OFS, called RAFTILOSE® is a mixture of ((glucosyl)l-fructosyl-fructose)n (GFn) oligomers (64%) and homoligomers (fructosyl-fructose)m (Fm) (36%), with a mean degree of polymerization of 4.8. The β-l,2 link between fructosyl moities makes OFS resistant to digestion in the upper gastrointestinal tract. Consequently, it enters the large bowel where it is largely fermented and induces proliferation of selected anaerobic bacteria, mostly bifidobacteria.[1] Therefore, OFS fulfills the criteria of a prebiotic with functional properties.[2]

Besides its effects on the gastrointestinal tract, OFS also induces systemic effects which may be beneficial for health. This paper aims at discussing the physiological effects of OFS by focusing more on lipid homeostasis.

2 RESULTS

Our results were obtained by analyzing the systemic effects of dietary OFS (RAFTILOSE®l0%) on lipid metabolism in male Wistar rats.

Table 1 shows the results obtained by comparing the influence of OFS given at the final concentration of 10% in basal diet of three types: either a standard diet (A04), a fibre-free diet (semi-synthetic diet without any digestible compound, except OFS), or a high fat diet known to induce hypertriglyceridemia and hypercholesterolemia. In all these protocols, the constant effect was a decrease in triglyceride (TG) concentration[3-5] This TG-lowering effect was significant (p < 0.05) after one week of feeding and lasted for up to 16 weeks (p <0.05), and was as significant in the post-prandial as in the post-absorptive period.[6]

Analysis of serum lipoproteins, after ultracentrifugation of the serum collected from fasted rats, showed that OFS feeding affects only the VLDL fraction while the composition of both low density lipoprotein (LDL) and high density lipoprotein (HDL) remained unchanged (Figure 1, adapted from reference 6).

Since a positive relationship exists between very low density lipoprotein (VLDL)-TG hepatic output and liver lipogenesis, we hypothesized that OFS could affect VLDL-TG secretion by reducing *de novo* fatty acid synthesis in the liver.

This hypothesis has been first investigated by measuring incorporation of [14]C- acetate (a lipogenic precursor) into TG in hepatocytes isolated from control and OFS-fed rats. OFS feeding significantly reduced (by more than two-fold) TG synthesis and secretion from acetate in isolated hepatocytes (Figure 2, adapted from reference 4). Our results indicate that the decreased hepatic lipogenesis occurs through a reduction in fatty acid synthase (FAS) activity, a key lipogenic enzyme.[4] Clarke and Jump have previously suggested that at least part of the hypotriglyceridemic action of dietary polyunsaturated fats is due to the inhibition of hepatic *de novo* fatty acid synthesis.[7]

In a third experiment, we demonstrated that OFS feeding, through its modulation of *de novo* lipogenesis, was able to protect rats against liver lipid accumulation but could not prevent the hypertriglyceridemia induced by a high load of fructose.[5]

Table 1 *Effects of 10% OFS on Post-Prandial Triglyceridemia in Rats*

Diet	Duration of the experiment (weeks)	% of serum TG as compared to controls
Standard diet	4	61
Fibre-free diet	5	27
High-fat diet	3	43

n≥5

Standard diet composition: carbohydrates 70.4%, including starch 38% - proteins 19.3% - lipids 3% - vitamin mix 1.3 % - mineral mix 6% (A04, UAR, Villemoisson-sur-Orge, France).

Fibre free diet composition: starch 32% - saccharose 32% - casein 20% - corn oil 8% - vitamin mix 2% - mineral mix 4% - methionine 0.2% (INRA, Jouy-en-Josas, France).

High-fat diet composition : starch 46%, saccharose 10%, non-digestible carbohydrates 5% - proteins 23% - lard 10% - corn oil 4% - cholesterol 0.15% - vitamin mix 0.8% - mineral mix 1.2% (INRA, Jouy-en-Josas, France).

OFS: Oligofructose; TG:Triglyceride

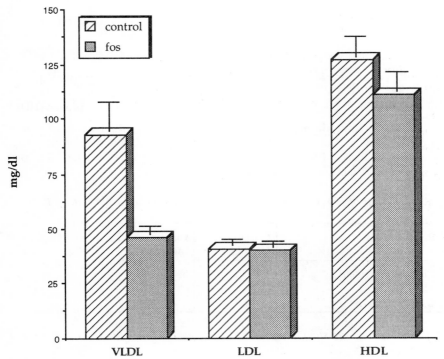

Figure 1 *Total lipoprotein content in the serum of control and fos-treated rats—fasted for 24 hours*
fos:fructooligosaccharides; VLDL:very low density lipoprotein; HDL: High density lipoprotein. Adapted from reference 6.

One must still explain how a non-digestible carbohydrate can regulate systemic lipid metabolism if, as shown in previous studies,[6] it has no significant effect on the fecal excretion of lipids.

By analogy with fermentable dietary fibres like pectin,[8] and knowing their effect on the physiology of the gastrointestinal tract, it could be possible that dietary OFS modifies the kinetics of absorption of dietary carbohydrates, leading to modifications of both serum glucose and hormones (insulin, glucagon). Moreover, dietary modulation of lipogenic activity is often linked to modifications of glucose and/or insulin serum levels: indeed, acarbose, an intestinal glucosidase inhibitor that delays starch digestion, reduces glucose absorption and post-prandial glycemia and insulinemia, also decreases FAS activity in the liver.[9] Similarly, resistant starch decreases serum TG concentration in rats,[10] reduces FAS activity by 50 and 20% in adipose tissue and liver respectively, and concomitantly lowers postprandial insulin response by 30% without affecting glucose response.[11]

Our data show that OFS ingestion reduces postprandial glycemia and insulinemia by 17 and 26%, respectively (Table 2, adapted from reference 4). This could thus explain the lower FAS activity, an enzyme for which transcription is primarily activated by glucose and insulin (for review see reference 12).

Studies are now in progress to analyse the putative involvement of glucose and insulin homeostasis in the hypolipidemic effect of OFS.

3 CONCLUSIONS

In our experimental studies in rats, we have shown that the main systemic effect of OFS is a decrease in serum TG. If the impact of dietary OFS or similar nutrients on serum lipids is confirmed in man, it could be interesting for human health, as hypertriglyceridemia is a risk factor for coronary heart disease (for review see reference 13).

In addition, through its modulation of *de novo* lipogenesis, OFS feeding is also able to protect against toxic effects, namely liver lipid accumulation, induced by a high concommitant intake of fructose. Such an interaction could be relevant if OFS is used as sugar or a fat substitute in a high fructose diet.

Table 2 *Effects of 10% OFS on Post-Prandial Glycemia and Insulinemia in Rats*

Diet	Glucose (mmol/L)	Insulin (ng/ml)
Control rats (n=10)	10.13±0.38	5.44±0.49
OFS-fed rats (n=10)	8.44±0.35**	3.67±0.49*

* and ** indicate statistically significant difference (Student's t test) for p <0.05 and 0.01, respectively. Adapted from reference 4.
OFS: Oligofructose

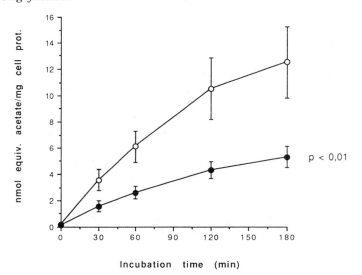

Figure 2 *Effect of oligofructose (OFS) feeding on triacylglycerol synthesis from* [14]*C-acetate*

Hepatocytes from control (open circles) or oligofructose-fed rats (closed circles) were incubated for 180 minutes in the presence of 2mM [1-[14]*C] acetate (0.2 mC1/mmol). Each point represents the mean ± SEM of 7 or 5 cell preparations for control and oligofructose-fed rats respectively. Adapted from reference 4.*

4 ACKNOWLEDGEMENTS

This study was supported by a grant from the European Community (CE DGXII, Contract N^0 AIR 2 -CT94-1095).
We thank "Raffineries Tirlemontoises", Belgium, for providing Raftilose®

5 REFERENCES

1. X.Wang and G.R.Gibson, Effects of the *in vitro* fermentation of oligofructose and inulin by bacteria growing in the human large intestine. *J. Appl. Bacteriol.* 1993, **75** : 373-380.
2. G.R.Gibson and M.B.Roberfroid, Dietary modulation of the human colonic microbiota: introducing the concept of prebiotics. *J. Nutr.* 1995, **125**: 1401-1412.
3. N.Delzenne, N.Kok, M.F. Fiordaliso, D. Deboyser, F. Goethals and M.Roberfroid, Dietary fructooligosaccharides modify lipid metabolism in the rat. *Am.J.Clin.Nutr.* 1993, **57**: 820S.
4. N.Kok, M. Roberiroid, A. Robert and N. Delzenne, Involvement of lipogenesis in the lower VLDL secretion induced by oligofructose in rats. *Br.J.Nutr.* 1996, **76**: 881-890.
5. N.Kok, M.Roberfroid and N.Delzenne. Dietary OFS modifies the impact of fructose on hepatic triacylglycerol metabolism. *Metabolism* 1996, **45**: 1547-1550.
6. M.Fiordaliso, N. Kok, J.P. Desager, F.Goethals, D.Deboyser, M.Roberiroid and N. Delzenne, Oligofructose-supplemented diet lowers serum and VLDL concentrations of triglycerides, phospholipids and cholesterol in rats. *Lipids*, 1995, **30**: 163-167.
7. S.D.Clarke and D.B.Jump, Dietary polyunsaturated fatty acid regulation of gene transcription. *Ann.Rev.Nutr.* 1994, **14**: 83-98.
8. M.B.Roberfroid, Dietary fibre, inulin and oligofructose: a review comparing their physiological effects. *Crit.Rev.Food Sci. Technol.* 1993, **33**: 103-148.
9. J.Maury, T. Issad, D.Perdereau, B.Gouhot, P.Ferré and J.Girard, Effect of acarbose on glucose homeostasis, lipogenesis and lipogenic enzyme gene expression in adipose tissue of weaned rats. *Diabetologia* 1993, **36**: 503-509.
10. E.A. de Deckere, W.Kloots and J.M.van amelsvoort, Both raw and retrograded starch decrease serum triacylglycerol concentration and fat accretion in the rat. *Br. J. nutr.* 1995, **73**: 287-298.
11. S.Takase, T.Goda and M.Watanabe, Monostearoylglycerol-starch complex its digestibility and effects on glycemic and lipogenic responses. *J. Nutr.Sci.Vitaminol.* 1994, **40**: 23-36.
12. F.D.Hillgartner, L.M.Salati and A.G.Goodridge, Physiological and molecular mechanisms involved in nutritional regulation of fatty acid synthesis. *Physiol.Rev.* 1995, **75**: 47-76.
13. S.M.Grundy, and M.A.Denke, Dietary influences on serum lipids and lipoproteins. *J. Lipid Res.* 1990, **31**: 1149-1172.

ACACIA GUM - A NATURAL SOLUBLE DIETARY FIBRE

T.P. Kravtchenko

Colloides Naturels International
129 chemin de Croisset
B.P. 4151
F-76723 Rouen
France

1 INTRODUCTION

'Let your food be your first medicine' (Hippocrates, 377 BC). This old aphorism has recently come back into fashion and has led to the development by the food industry of the so-called nutraceuticals or functional foods, designed for actively promoting health. Since the beginning of the 1970s, the low intake of dietary fibre in Western countries has been of increasing concern for consumers. Foods fortified with fibre were developed, long before the term functional food existed. More than a pure marketing concept, the development of these 'new' foods meets the demands of the modern consumer.

However, though consumers are more aware of healthy eating, they do not want to eat without pleasure: fibre-fortified foods must have the same organoleptic qualities as conventional foods. In practice, the addition of significant amounts of dietary fibre is often limited by side effects such as sandy texture, increase in viscosity, or bad taste or odour. Acacia gum offers the possibility of being added in large amounts while maintaining the original taste and texture of the food in which it is incorporated.

2 ACACIA GUM - A NATURAL FOOD INGREDIENT

Acacia gum, also known as gum arabic, is defined in the Pharmacopoeias as 'the gummy exudates flowing naturally or obtained by incision of the trunk and branches of Acacia senegal and other species of African origin'. There are many Acacia species (some 700) of which few can provide gum volumes of industrial interest. After harvest, which is carried out during the dry season (December-May in the sub-Saharian area), acacia gum is purified (Figure 1) by a simple physical process (centrifugation and filtration) without chemical or enzymatic modification. A simple sterilisation procedure ensures bacterial safety. Finally, highly water-soluble powders are obtained by spray-drying and/or granulation.

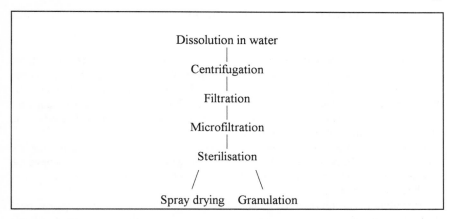

Dissolution in water
|
Centrifugation
|
Filtration
|
Microfiltration
|
Sterilisation
/   \
Spray drying   Granulation

Figure 1 *Purification process of acacia gum*

From a chemical point of view, acacia gum is made of a galactan main chain carrying heavily branched galactose/arabinose side-chains. Some rhamnose and/or glucuronic acid may be present as side-chain terminations (Figure 2). The overall chemical composition may vary slightly from one Acacia species to another or in relation to its geographical origin.

From a regulatory point of view, acacia gum is recognised as a food additive by the joint FAO/WHO committee on food additives (JECFA) and by the EU, with no ADI (acceptable daily intake) specified. In the USA, the FDA has recognised acacia gum as GRAS (generally recognised as safe) for many years and it is registered in the US Food Chemical Codex under reference 1841330. In the EU, acacia gum is also compatible with foods conforming to the organic mode of production (Reg. 2092/91).

```
                                    GAL
                                     |
                                    ARA
                                     |
                                    GAL          ARA
                                     |            |
           MeGLcA — GAL —  GAL       GAL
                                     |            |
        ARA                         GAL          ARA
         |                           |            |
        ARA                         GAL          GAL      GLcA — RHA
         |                           |            |                 |
        ARA               ARA       GAL          ARA              ARA — GAL
         |                 |         |            |                 |
    ·····GAL — GAL —  GAL —  GAL —  GAL — GAL — GAL ·····
         |         |         |       |            |       |
    GAL — ARA     GAL       ARA      ARA         GAL      ARA
               |       GLcA            |            GLcA
             GLcA       |            GAL             |
               |       RHA   ARA — GAL             RHA
             RHA                     |
                                    ARA
                                     |
                                    GAL
```

Figure 2 *Hypothetical chemical structure of acacia gum*
Source: Street and Anserson, 1983.[1] ARA: arabinose; GAL: galactose; GlcA: glucuronic acid; RHA: rhamnose;

3 THE CONCEPT OF DIETARY FIBRE

Dietary fibre is a general term covering a wide variety of substances that are not digested in the upper part of the human digestive tract. Several definitions have been proposed but one of the most relevant is that proposed by Trowell and Burkitt[2] and retained by the Association of Official Analytical Chemists (AOAC): 'the remnants of plant cells resistant to hydrolysis by the alimentary enzymes of man. It is composed of cellulose, hemicellulose, oligosaccharides, pectins, gums, waxes and lignin'.

The passage of dietary fibre through the digestive tract results in several physiological effects which may be profitable for human health. However, all fibres do not behave the same way. First of all, one may differentiate fibres which are soluble in water from those which remain insoluble.

Insoluble fibres, represented by cereal brans and cell-wall residues (eg soy, pea) mainly play a mechanical role, increasing transit time and affecting the digestion of lipids by adsorbing the biliary salts needed for emulsifying fat globules. Because of their hydrophobicity, insoluble fibres also prevent colon cancers by absorbing hydrophobic carcinogens such as pyrene derivatives or heterocyclic aromatic amines. Because of their insolubility they are poorly fermented by the colonic flora and are thus almost not metabolised.

In contrast, soluble fibres are characterised by the fact that they are degraded by the bacteria present in the large bowel, resulting in several metabolic changes:

• lowering of the luminal pH
• stimulation of the endogenous flora
• production of short chain fatty acids (SCFA).

The extent of each of these effects depends on the chemical nature of the fibre considered.

Soluble fibre sources may further be divided into high- and low-viscosity fibres. The former type affect the transit as well as the digestion of nutrients because of their effect on the viscosity of the intestinal content. The latter type, to which acacia gum belongs, mainly affects human physiology by systemic effects via the intestinal flora and its production of SCFA.

4 NUTRITIONAL PROPERTIES OF ACACIA GUM

4.1. Mechanical Contribution to Transit

Compared with insoluble or highly viscous fibres, acacia gum only plays a limited role in the upper gastrointestinal tract, probably because of its low viscosity.[3]

In contrast to other low-viscosity fibres, such as oligosaccharides, acacia gum does not exhibit laxative side effects. Large amounts of undigested low-molecular weight (MW) molecules increase the intraluminal osmotic pressure, stimulating the migration of water from the body to the intestinal contents and therefore provoking diarrhoea by excess water. Thanks to its polymeric nature (MW higher than 300,000 daltons) acacia gum does not disturb the osmotic pressure and thus does not exhibit laxative side effects.

Clinical tests with human subjects showed good tolerance towards acacia gum at intakes up to 50g/day.[4]

4.2 Prebiotic Effects

In the large bowel acacia gum is readily fermented by the endogenous bacterial flora. McLean-Ross *et al.* showed that no acacia gum could be recovered from the feces of humans[5] or rats[6] fed acacia gum, thus indicating that it is completely fermented by the human colonic flora. Indeed, fermentable acacia gum promotes the growth of resident gut microbes; however, such a fermentation requires an adaptation period of a few days.[7,8]

The human colon contains a complex bacterial ecosystem where at least 400 distinct species coexist in equilibrium with their human host.[9] Dietary fibre represents an important source of carbon, and enrichment of the diet with acacia gum promotes the increase of the number of excreted bacterial cells,[4,10,11] thus increasing stool weight and facilitating expulsion of the feces.

However, among the several hundreds of different bacterial species, those belonging to the genus Bacteroides, Bifidobacterium and Lactobacillus are particularly stimulated. Wyatt *et al.*[8] observed that the proportion of the fecal flora able to degrade acacia gum (ie Bifidobacterium, Bacteroides and Lactobacillus genera) rose from an initial level of 6.5% to more than 50% during acacia gum supplementation. This effect is called bifidogenic or prebiotic,[12] ie growth stimulation of the endogenous beneficial bacterial strains. This change in the species composition of the colonic microflora may be explained by the fact that not all bacterial species possess the necessary enzymes to degrade fibre molecules,[13] but other reasons such as the change of pH may be taken into account. The beneficial effects of the intestinal flora have been described extensively elsewhere[14] and will not be discussed in this review.

A large part of the metabolic effects of soluble dietary fibre may be ascribed to its fermentation by the colonic flora.[14,15]

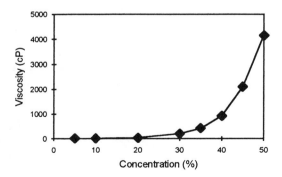

Figure 3 *Influence of concentration on the viscosity of an acacia gum solution*

Table 1 *Effects of Acacia Gum on the Stool Output in Humans*

	Control	After three weeks acacia gum consumption
Wet stool weight (g/day)	147	161
Dry stool weight (g/day)	37	52
Total fecal SFCA (mmol/day)	13.3	11.5
Bile acids (mmol/day)	1.18	1.06
Transit time (hours)	51	70

Source: McLean-Ross et al., 1983.[5]

4.3 Gas Production

The fermentation of acacia gum leads to the formation of hydrogen, carbon dioxide, and methane. [5,6,11] These gases constitute final metabolites which are normally expelled with the feces or on the breath. However, in the presence of large amounts of highly fermentable fibre, fermentation is rapid in the cecum and gas production may provoke flatulence.[16] Because of its highly branched chemical structure which is hardly degraded by bacterial enzymes, acacia gum fermentation is slow. Gas production is therefore delayed and displaced all along the large gut without provoking the feeling of bloating.

4.4 Production of Short Chain Fatty Acids

The fermentation of acacia gum also leads to the formation of SCFA, as has been shown in rats, [6,7,10,11,17-20] *in vitro* tests using human fecal inoculates[21-24] and also in human subjects.[5]

The first observable effect of SCFA production is a drop of the intraluminal pH[7,18,22] which is usually considered as a beneficial factor for the health of the colon. Lowering the intraluminal pH also contributes to modulating distribution of the various bacterial

Table 2 *Change in the Human Fecal Flora during Feeding 10g Acacia Gum/day*

	Total bacteria	Bacteria-fermenting acacia gum	Percentage fermenting acacia gum
Initial control	4.3×10^{11}	1.6×10^{10}	6.5 ± 3.1
Treatment : days 2-5 days 9-19	2.6×10^{11} 6.3×10^{11}	4.8×10^{10} 3.5×10^{11}	18.3 ± 2.8 53.6 ± 9.7
Recovery : days 7-9 days 45-48	6.8×10^{11} 3.6×10^{11}	2.7×10^{11} 1.8×10^{10}	34.5 ± 5.6 5.1 ± 0.0

Source: Wyatt *et al.* 1986.[8]

species (prebiotic effect).

SCFA are efficiently absorbed from the colonic lumen and are metabolised by various tissues. Butyrate is preferably metabolised by colonocytes, propionate is almost quantitatively taken up by the liver and may modulate hepatic carbohydrate and lipid metabolism[25]whereas acetate largely escapes colonic and hepatic metabolism and probably serves primarily as a fuel for peripheral tissues such as muscles. Compared with other fibre sources, acacia gum elicits the formation of butyrate and propionate at the expense of acetate, thus promoting several metabolic effects. A high proportion of butyrate and propionate may be due to the complexity of the acacia gum molecules.[26]

There is some evidence that acacia gum stimulates intestinal epithelial cell turnover,[27] thus contributing to the maintenance of the mucosal wall. Several studies on rats[7,10,11,20] showed that acacia gum also elicits a marked enlargement of the cecal lumen and a trophic effect on the cecal wall in rats.

Dietary fibre has often been shown to reduce blood lipids. In rats fed hypercholesterolemic diets and acacia gum, Annison *et al.*[17] did not find any effect on the concentration of plasma and liver cholesterol, whereas Kiriyama *et al.*[28] found an appreciable depressing activity on cholesterol levels. In man, Haskell *et al.*[29] and Jensen *et al.*[30] did not find any effect of acacia gum on any plasma lipid parameter. By feeding higher quantities of acacia gum (at least 25g/day), McLean-Ross *et al.*,[5] Sharma[31] and Eastwood *et al.*[32] observed a significant decrease of the plasma cholesterol concentration, especially low density lipoprotein and very low density lipoprotein fractions. As compared with other hypocholesterolemic fibres, acacia gum has almost no effect on the viscosity of the intestinal content and does not alter fecal bile acid excretion.[5,10,32] Moreover, the hypocholesterolemic effect of acacia gum requires a diet adaptation period of a few days, reinforcing the idea that it is mediated by the SCFA produced by the colonic microflora.

Feeding acacia gum increases urea flux from blood to cecum and increases net nitrogen retention in the cecum, resulting in a significant decrease of the blood urea content.[4,20,23,33]

Modulation of the production of different SCFA by feeding acacia gum thus has potential health benefits. However, because of the complexity of the mechanisms

Table 3 *Fermentation of Acacia Gum in the Rat*

	Control	After 12 weeks acacia gum consumption
Dry cecal content (g)	0.47±0.09	1.19±0.07
Total cecal SCFA (mol)	276±42	622±102
Total fecal SCFA (mol/day)	109±40	102±14
Individual cecal SCFA (%) : **Acetate** **Propionate** **Butyrate**	70.8±4.0 15.5±0.9 8.4±0.5	55.7±3.0 23.0±8.6 12.5±1.1

Source: Walter *et al.*, 1988.[11]

Table 4 *Effects of Acacia Gum (25g/day) on Human Blood Lipids*

	Control	**After three weeks acacia gum consumption**
Phospholipids	264±20.8	270±25.5
Triglycerides	1.64±0.61	1.46±0.49
Total Cholesterol	6.22±0.54*	5.83±0.65*

*p<0.05, Wilcoxon rank sum test
Source: Mc Lean-Ross *et al.*, 1983.[6]

involved, many conclusions relative to acacia gum remain to be confirmed. It is the reason why Colloides Naturels International (CNI) still actively promotes academic research on the nutritional properties of acacia gum.

5 APPLICATIONS OF ACACIA GUM AS A DIETARY FIBRE IN FOOD PRODUCTS

Due to its low viscosity and its absence of taste and odour, acacia gum can be added in large quantities without disturbing the organoleptic properties of the food product in which it is incorporated. In order to guarantee a high technological neutrality, CNI has developed a new range of selected acacia gums called FIBREGUM.

Traditional confectionery gums may contain more than 40% acacia gum. Other conventional food products, processed ingredients or even dietetic specialities have been formulated in order to compensate for the lack of dietary fibre in usual Western diets. These products usually contain 2-5% acacia gum, providing 2-15g dietary fibre/serving.

It must kept in mind that acacia gum is traditionally used as a food additive, thanks to its unique texturising (confectionery) or stabilising (beverages) properties, but also for numerous other applications.[34,35] It is thus possible to combine fibre-enrichment with a given functional property. In cereal bars FIBREGUM TX provides sticking properties as well as good moisture stability on storage. In instant powdered beverages, FIBREGUM INSTANT may be used to encapsulate natural flavours and/or vitamins to protect them against oxidation and minerals to limit their oxidising power. In dietetic tablets, FIBREGUM INSTANT serves as cohesive agent which can be used for direct compression. In salad dressings, FIBREGUM AS can replace conventional emulsifier/stabiliser combinations without affecting the overall product stability.

6 REFERENCES

1. C.A. Street and D.M.W. Anderson, *Talanta*, 1983, **30**(11), 887.
2. H. Trowell and D. Burkitt, *Bol. Assoc. Med. P. Rico*, **dec**. 1986, 541.
3. J. Adiotomre, M.A. Eastwood, C.A. Edwards and W.G. Brydon, *Am. J. Clin. Nutr.*, 1990, **52**, 128.
4. D.Z. Bliss, T.P. Stein, C.R. Schleifer and R.G. Settle, *Am. J. Clin. Nutr.*, 1996, **63**, 392.
5. A.H. McLean-Ross, M.A. Eastwood, J.R. Anderson and D.M.W. Anderson, *Am. J. Clin. Nutr.*, 1983, **37**, 368.
6. A.H. McLean-Ross, M.A. Eastwood, W.G. Brydon, A. Busuttil, L.F. Mc Kay and D.M.W. Anderson, *Brit. J. Nutr.*, 1984, **51**, 47.
7. B. Tulung, C. Rémésy and C. Demigné *C, J. Nutr.*, 1987, **117**, 1556.
8. G.M. Wyatt, C.E. Bayliss and J.D. Holcroft, *Brit. J. Nutr.*, 1986, **55**, 261.
9. N. Gournier-Chateau, 'Les probiotiques en alimentation animale et humaine', Technique et Documentation, Lavoisier, Paris, 1994, p. 9.
10. D.J. Walter, M.A. Eastwood, W.G. Brydon and R.A. Elton, *Brit. J. Nutr.*, 1986, **55**, 465.
11. D.J. Walter, M.A. Eastwood, W.G. Brydon and R.A. Elton, *Brit. J. Nutr.*, 1988, **60**, 225.
12. G.R. Gibson and M.B. Roberfroid, *J. Nutr.*, 1995, **125**, 1401.
13. A.A. Salyers, A.P. Kuritza and R.E. McCarthy, *Proc. Soc. Exp. Biol. Med.*, 1985, **180**, 415.
14. M.B. Roberfroid, F. Bornet, C. Bouley and J.H. Cummings, *Nutr. Rev.*, 1995, **53**(5), 127.
15. T. Hoverstad *,J. Scand. Gastroenterol.*, 1986, **19** (suppl. 93), 89.
16. J. Stevens, D.A. Levitsky, P.J. Van Soest, J.B. Robertson, K.J. Kalkwarf and D.A. Roe, *Am. J. Clin. Nutr.*, 1987, **46**, 812.
17. G. Annison, R.P. Trimble and D.L. Topping, *J. Nutr.*, 1995, **125**(2), 283.
18. G.B. Storer, R.J. Illman, R.P. Trimble, A.M.P. Snoswell and D.L. Topping, *Nutr. Res.*, 1984, **4**, 701.
19. D.L. Topping, R.J. Illman and R.P. Trimble, *Nutr. Rep. Int.*, 1985, **32**(4), 809.
20. H. Younes, K.A. Garleb, S.R. Behr, C. Rémésy and C. Demigné, *FASEB J.*, 1994, **8**, A186.
21. L.D. Bourquin, E.C. Titmeyer, G.C. Fahey and K.A. Garleb, *J. Scand. Gastroenterol.*, 1993, **28**(3), 249.
22. T. May, R.I. Mackie, G.C. Fahey, J.C. Cremin and K.A. Garleb, *J. Scand. Gastroenterol.*, 1994, **29**(10), 916.
23. P.B. Mortensen, H. Hove, M.R. Clausen, K. Holtung, Scand. *J. Gastroenterol.*, 1991, **26**, 1285.
24. E.C. Titmeyer, L.D. Bourquin, G.C. Fahey and K.A. Garleb, *Am. J. Clin. Nutr.*, 1991, **53**, 1418.
25. C. Demigné and C. Rémésy, 'Short-chain fatty acids: metabolism and clinical importance', A.F. Roche ed., Report of the tenth Ross conference on medical research, Columbus, Ohio, Ross Laboratories, 1991, p. 17.
26. C. Rémésy, C. Demigné and F. Chartier, *Reprod. Nutr. Develop.*, 1980, 20, 1339.

27. M.D. Howard, D.T. Gordon, K.A. Garleb and M.S. Kerley, *J. Nutr.*, 1995, **125**, 2604.
28. S. Kiriyama, Y. Okazaki and A. Yoshida, *J. Nutr.*, 1969, **97**, 382.
29. W.L. Haskell, G.A. Spiller, C.D. Jensen, B.K. Ellis and J.E. Gates, *Am. J. Cardiol.*, 1992, **69**(5), 433.
30. C.D. Jensen, G.A. Spiller, J.E. Gates, A.F. Miller and J.H. Whittam, *J. Am. Coll. Nutr.*, 1993, **12**(2), 147.
31. R.D. Sharma, *Nutr. Res.*, 1985, **5**, 1321.
32. M.A. Eastwood, W.G. Brydon and D.M.W. Anderson, *Am. J. Clin. Nutr.*, 1986, **44**, 51.
33. S.A. Assimon and T.P. Stein, *Nutr.*, 1994, **10**(6), 544.
34. A. Imeson, 'Thickening and gelling agents for food', Blackie Academic & Professional, London, 1992, p. 66.
27. F. Thevenet, 'Encapsulation and controlled released of food ingredients', Risch S.J. & Reineccius G.A. eds, ACS Symposium Series, American Chemical Society, Washington DC, 1995, p. 51.

IMMUNOMODULATING PROPERTIES OF A STRAIN OF *BIFIDOBACTERIUM* USED AS PROBIOTIC ON THE FECAL AND CELLULAR INTESTINAL IgA ANTIROTAVIRUS RESPONSES IN MICE

M.C. Moreau, N. Bisetti and C. Dubuquoy

UEPSD
Bât 440
INRA
78352 Jouy-en Josas
Cedex, France.

1 INTRODUCTION

It has been claimed that fermented milks could have beneficial effects on human health, especially on the reduction or prevention of diarrhoea in infants.[1-4] Previous studies[3,5] have suggested that a possible mechanism underlying this effect could be the immunostimulation of intestinal IgA response by live lactic acid-producing bacteria ie *Lactobacillus, Streptococcus* and *Bifidobacterium,* which are present in high numbers in fermented milks. Such bacteria are termed probiotics.

The aim of this work was to evaluate the immunostimulating properties of *Bifidobacterium* DN-173010, used as probiotic in fermented milks, on the intestinal IgA anti-rotavirus antibody (Ab) responses, measured in feces and at cellular level in the lamina propria of the small intestine. Adult gnotobiotic mice harbouring only the *Bifidobacterium* DN-173010 strain in the gut were infected with a heterologous simian rotavirus strain (SA-11), and the intestinal IgA anti-rotavirus response was compared with that of germ-free mice. This experimental animal model provided evidence on the adjuvant effect of a *Bifidobacterium* strain on the enhancement of IgA anti-rotavirus Ab.

2 MATERIAL AND METHODS

2.1 Animals and Viral Infection

Germ-free (GF) adult C3H/He male and female mice, eight weeks old, were housed in a Trexler plastic isolator. They were infected with a heterologous simian rotavirus strain (SA-11) kindly provided by J. Cohen (INRA, VIM, 78350 Jouy-en-Josas, France). The virus was grown in MA104 cells in a complete minimal essential medium supplemented with fetal calf serum, released by freezing and thawing, and stored at -80°C.[6] The dose of virus administered in 500µl tissue culture supernatant by stomach probe was 3.10^8 PFU. Gnotobiotic mice harbouring *Bifidobacterium* in the gut were infected with rotavirus three weeks after bacterial implantation.

2.2 Bacteriological Assays

Bifidobacterium DN 173010 (from DANONE collection) was given by "Le Centre International de Recherche Daniel Carasso" (CIRDC), Le Plessis Robinson, France. It was grown on Mann-Rogosa Sharp medium (MRS, Difco) with 0.03% of L-cysteine, for 16 hours at 37°C. Mice were inoculated with a 0.4ml dose of a fresh culture by stomach probe (ie 10^8 bacteria). Bacterial implantation was verified by bacterial counts performed on fresh feces once a week.

2.3 Fecal Collection and Detection of Anti-rotavirus IgA and Total IgA in Feces

Individual stool samples collected were stored at -20°C. At the time of innoculation, a 10% suspension was prepared using stool diluent containing Pepstatin A (Sigma) 1μg/ml, Aprotinin (Sigma) 1,25μg/ml, EDTA 0.2mM, phenylmethylsulphonyl fluoride 3mM, 10 mM sodium azide and 5% fetal calf serum. After homogenisation and centrifugation (650g, 10 minutes) the levels of anti-rotavirus IgA Ab and total IgA in supernatants were measured by ELISA. Briefly, microtitre plates (Immunlon II, Dynatech) were coated with a rabbit anti-rotavirus hyperimmune serum overnight at 4°C. Free sites were blocked with 1% bovine serum albumin for 1hour at 37°C. Plates were coated with stock virus diluted in phosphate-buffered saline (PBS) and incubated for 1hour at 37°C. Intestinal samples (100μl) were serially diluted two-fold in PBS-Tween and incubated for 1.5hours at 37°C. A goat anti-mouse IgA horseradish peroxydase conjugate (Sigma) was added. After incubation (1.5hours, 37°C), 100μl substrate (Tetramethylbenzidine; Sigma) was added and the reaction stopped by the addition of 50μl H_2SO_4 4N. Absorbance was measured at 450nm (Multiskan reader, Labsystem). At each step of the process, the plates were washed with PBS containing 0.05% Tween 20. The IgA anti-rotavirus antibody titer was expressed in arbitrary units (AU). A standard fecal extract arbitrarily assigned to contain one AU/g feces was used to construct a standard curve for each plate, and units of anti-rotavirus/g feces were calculated by reference to this curve.

Measurement of total IgA differs by the coating which was performed with a sheep anti-mouse IgA (Sigma) overnight at 4°C. Intestinal samples were serially diluted three-fold. The following steps were the same as above. Results are expressed in μg IgA/g feces from a standard curve obtained with a commercial IgA serum (ICN).

2.4 Determination of the Number of Anti-Rotavirus IgA- and Total IgA-Secreting Cells by ELISPOT

Lamina propria lymphocytes from GF and gnotobiotic mice were extracted 30 days after rotavirus innoculation as described elsewhere.[7] Anti-rotavirus-specific Ab-secreting cells (ASC) and total IgA-secreting cells (IgA-SC) were enumerated by ELISPOT.[8] Briefly, the first steps of coating were the same as described above for ELISA. Lymphoid cells suspended in RPMI 1640 (Flow laboratories) were added to the wells and serially diluted two-fold. After centrifugation (5 minutes, 1200g), the plates were incubated for 3hours 37°C in a CO_2 incubator. After extensive washing with water and PBS-Tween, a rabbit anti-mouse IgA alkaline phosphatase (Sigma) was used as the conjuguate, added to the wells and incubated overnight at 4°C. After washing, the wells were filled with a substrate-agarose mixture that consisted of 2.3 mM 5-Bromo-4-chloro-3-indolyl phosphate (Sigma), 0.1M Amino-2-methyl-1-propanol buffer and 0.6% agarose. After 1hour at room temperature, spots were visible and counted using a reverse microscope (Nikon).

2.5 Statistical Analyses

Student's T test was used for comparison between two means.

3 RESULTS

3.1 Adjuvant Effect of *Bifidobacterium* DN 173010 on the Fecal IgA Anti-Rotavirus Response

The GF adult mouse model is convenient to study the active intestinal immunization following viral infection with a heterologous rotavirus SA-11. Anti-rotavirus-specific IgA Ab are detected in feces from 16 days post-infection with a peak at 22 days (Figure 1, top), and rotavirus excretion was measured by three days after inoculation (data not shown). *Bifidobacterium* DN-13010 was established in high numbers in the gut of gnotobiotic mice (7×10^9/g feces). Gnotobiotic mice exhibited a different kinetics (Figure 1, bottom). The Ab response was detected earlier ie 12 days post-infection and reached a peak at 18 days. The level was two-fold higher than in the GF group (P 0.01). In contrast total IgA levels/g feces were similar in two groups (0.51±0.05 in GF mice versus 0.53±0.06 in gnotobiotic mice).

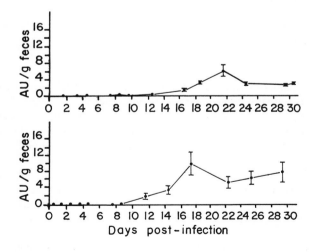

Figure 1 *Kinetics of the anti-rotavirus IgA antibody response in feces of GF mice (top) and gnotobiotic mice harbouring Bifidobacterium DN-173010 in the gut (bottom). Each point is expressed as mean of antibody titers ± SEM of groups of eight to 10 mice*

Table 1 *IgA-secreting cell (IgA-SC) and rotavirus antibody IgA secreting-cell (ASC) numbers in the lamina propria of GF and gnotobiotic mice harbouring Bifidobacterium DN-173010 in the gut. Results are expressed as mean of secreting-cell numbers ± SEM.* ** P 0.01

Groups of mice	IgA-SC/10^6 cells	ASC/10^6 cells	Ratio of ASC/ IgA-SC x 100
Germ-free (n=9)	17 000±8 000	447±198	3.80±1.74
Gnotobiotic mice (n=8)	33 000±10 000	5486±2670**	12.5±5.65**

3.2 Adjuvant Effect of *Bifidobacterium* 173010 on the Number of Anti-Rotavirus IgA-Secreting Cells Present in the Lamina Propria

The number of rotavirus Ab IgA secreting-cell (ASC) was clearly greater in gnotobiotic mice than in GF mice (Table 1). This result confirms the immunostimulating effect of *Bifidobacterium* measured in feces, and indicates that the adjuvant effect is expressed at the cellular level in the lamina propria. No enhancing effect of *Bifidobacterium* on the number of IgA-SC was observed.

4 DISCUSSION

The experimental model of adult GF mice infected with a heterologous rotavirus allowed us to investigate the immunomodulating properties of a commercial strain of lactic-acid producing bacteria, *Bifidobacterium* DN 173010 on the enhancement of the intestinal anti-rotavirus IgA antibody response. This effect was found both at the cellular and fecal levels.

In this work we show that an infection with a heterologous rotavirus in adult GF mouse was capable of inducing active intestinal IgA anti-rotavirus Ab response. We know that the digestive microflora is a environmental factor which has a profound influence on the gut-associated lymphoid tissue.[9] This mouse model allowed us to study the modulating effect of a bacteria present in the intestine on the anti-rotavirus Ab response.

These preliminary results suggest that fermented milks could have a beneficial effect on infant health by stimulation of the specific intestinal IgA immune response against enteric pathogens. *Bifidobacterium* is abundant in the gut flora of breast-fed babies who are more resistant to enteric infection than bottle-fed babies.[10] This bacteria is thought to play a beneficial role in breast-fed babies. On the other hand, fermented milks containing a high level of probiotics, which can exert a beneficial effect during transit time, could have a similar role to resident intestinal *Bifidobacterium*. Clinical studies have reported the protective effect of the supplementation of infant formula with fermented milks or probiotics, on the reduction of acute rotavirus diarrhoea duration and diarrheal frequency in infants.[1-4] Rotaviruses are recognized as the major cause of severe infantile diarrhoeas worldwide, and several studies have emphasized the importance of IgA Ab at the mucosal surface of the small intestine as a major determinant of resistance to rotavirus illness.[6] It would now be interesting to test the modulating effect of fermented milks, and other probiotics, on the anti-rotavirus intestinal IgA Ab response in the different conditions of intestinal bacterial environment which are found in babies, using the gnotobiotic model presented here.

According to our results, fecal specific IgA Ab could be a good biomarker to test the immunomodulating properties of a food component. Further studies are now needed substantiate this finding.

5 ACKNOWLEDGEMENTS

We thank Hervé Prévoteau for his excellent technical assistance, and the CIRDC (Groupe Danone) for the financial support to the N. Bisetti's work.

6 REFERENCES

1. G. Boudraa, M. Touhami, P. Pochart, R. Soltana, J.Y. Mary and J.F. Desjeux, *J. Pediatr. Gastroenterol. Nutr.*, 1990, **11**, 509.
2. S. Gonzalez, G. Albarracin, M. Locascio de Ruiz Pesce, M. Male, M.C. Apella, A. Pesce de Ruiz Holgado and G. Oliver, *Microbiol. Alim. Nutr.*, 1990, **8**, 349.
3. M. Kaila, E. Isolauri, E. Soppi, E. Virtanen, S. Laine and H. Arvilomi, *Pediatr. Res.*, 1992, **32**, 141.
4. J.M. Saavedra, N.A. Bauman, I. Oung, J.A Perman and R.H. Yolken, *The Lancet*, 1994, **344**, 1046.
5. G. Perdigon, S. Alvarez, M. Rachid, G. Agüero and N. Gobbato, *J. Dairy Sci.*, 1995, **78**, 1597
6. R.L. Ward, *J. Infec. Dis.*, 1996, **174** (Suppl), S51.
7. N. Lycke, *Scand. J. Immunol.*, 1986, **24**, 393.
8. A.T.J. Bianchi, J.W. Scholten, I.M.C.A. Jongenelen and G. Koch, *Vet. Immunol. Immunopath.*, 1990, **24**, 125.
9. M.C Moreau and M. Coste, *"World Rev Nutr Diet"*, A.P. Simopoulos, Basel, Karger, 1993, vol **74**, p 22.
10. M.C. Moreau, M. Thomasson, R. Ducluzeau et P. Raibaud, *Reprod. Nutr. Develop.*, 1986, **26**, 745.

EFFECT OF A RESISTANT STARCH ON BIOCHEMICAL AND MUCOSAL MARKERS OF COLONIC NEOPLASIA IN RATS.

C. Rumney, I. Rowland, G. Caderni,[1] P. Dolara,[1] B. Pool-Zobel,[2] G. Morozzi,[3] S. Silvi[4] and A. Cresci[4]

BIBRA International, Carshalton, Surrey, SM5 4DS
[1]Department of Pharmacology, University of Florence, 50134 Florence, Italy
[2]Bundesforschungsanstalt für Ernährung, 76131 Karlsruhe, Germany
[3]University of Perugia, Perugia, Italy;
[4]University of Camerino, 62032 Camerino, Italy

1 INTRODUCTION

There is evidence that the type of carbohydrate in the diet, particularly whether it is a simple sugar or a complex carbohydrate, can have marked effects on the colonic mucosa function, possibly inflencing colon cancer. We have studied the differential effects of sucrose, maize starch (a digestible starch) and CrystaLean (a retrograded amylomaize, and thus resistant, starch) on various mucosal and metabolic parameters associated with colonic cancer.

In order to circumvent the problem of differences in gut flora between laboratory animals and man, we have used human-flora-associated (HFA) rats in which germ-free rats are colonised with a complete human fecal microflora. These rats retain the bacteriological and metabolic characteristics of the human microflora.[1]

2 MATERIALS AND METHODS

2.1 Experiment 1

Thirty germ-free F344 rats were given 1ml of a fresh human fecal suspension (20% w/w) *p.o.* to produce HFA rats. One week later, the rats were divided into three groups of 10 and transferred onto one of three experimental diets which were modifications of AIN 76 in which the calcium concentration was reduced and were based on a diet previously used by Caderni *et al.*[2] The diets were as follows:

1. Sucrose (46% w/w) diet
2. Maize starch ('Globzeta' 46% w/w) diet
3. CrystaLean (31% Globzeta; 15% CrystaLean) diet

After four weeks on the diets, the animals were killed, the colons removed and a biopsy taken for measurement of mucosal proliferation by two methods, namely incorporation of ^{3}H thymidine and by microdissection of colonic crypts. Colon contents were removed for measurement of long chain fatty acids (LCFA) and bile acids (BA) and cecal contents were removed for measurement of short chain fatty acids (SCFA), β-glucuronidase activity, ammonia concentration and conversion of the cooked food mutagen IQ to its direct acting genotoxic metabolite, 7-OHIQ.[3]

Three parallel groups of 10 rats, treated as above, were given by gavage either saline (four rats/group) or the model colon carcinogen dimethylhydrazine (DMH; 15 mg/kg; six rats/group). Sixteen hours later, the rats were killed, the colons removed and DNA damage in the colonic mucosa assessed by the Comet assay.[4]

2.2 Experiment 2

Thirty-one germ-free F344 rats were dosed with 1ml of a fresh 10% (w/v) human fecal suspension to produce HFA rats. Fifteen rats were given a suspension prepared from pooling three fecal samples from UK volunteers, whilst the remaining sixteen rats were given a suspension prepared from pooling three fecal samples from Italian volunteers. After one week, the rats were further subdivided by transferring to a sucrose diet or CrystaLean diet (as above). After four weeks on the diet, the rats were killed and cecal contents removed for bacteriological analysis.

In both experiments, rats were maintained in germ-free isolators throughout the experimental period and were given water and diet previously sterilized by exposure to 5Mrad of γ-radiation from a ^{60}Co source.

Results were subjected to analysis of variance using the Minitab statistical package. Individual means were compared using the Least Significant Difference criterion.[5]

3 RESULTS

Cecal weight was increased by about 15% in the animals fed maize starch and by approximately 100% (P<0.001) in rats fed CrystaLean. By comparison with the sucrose-fed animals, cecal pH (6.71) was reduced (P<0.001) by both maize starch (6.31) and CrystaLean (6.13). Colonic pH (6.86 with sucrose) was also reduced by both maize starch (6.5; P<0.01) and CrystaLean (5.96; P<0.001), the reduction being greater with CrystaLean. Total SCFA levels did not differ significantly between the three dietary groups, although the proportions of acetate, propionate and butyrate were altered. Maize starch decreased propionate production from approximately 13.6% with sucrose to 10.2% (P<0.01), whilst CrystaLean decreased acetate from approximately 78.7% to 71.5% (P<0.05), decreased propionate to 6.7% (P<0.001) and increased butyrate production from approximately 2.5% to 13.7% (P<0.001). BA concentration in colon contents was not significantly altered by diet, but both maize starch and CrystaLean reduced (P<0.001) the concentration of colonic LCFA from 7.1 mg/g fecal dry weight with the sucrose diet to 2.6 and 1.8 mg/g respectively.

Cecal β-glucuronidase activity (120μmol p-nitrophenol produced/hour/g cecal contents with the sucrose diet) was significantly reduced (P<0.001) to 62 μmol/hour/g by both maize starch and CrystaLean. In rats on sucrose diet, cecal ammonia concentration (24μmol/g) was reduced by maize starch to 19μmol/g (P<0.05) and to a greater extent by CrystaLean to 5.2μmol/g (P<0.001). Both maize starch and CrystaLean reduced (P<0.001) the rate of formation of 7-OHIQ by cecal contents *in vitro*, from 12.7%/hour with the sucrose diet to 6%/hour and 3.8%/hour respectively.

Crystalean increased mucosal proliferation in the distal colon, as measured by the mean number of mitoses per crypt using the microdissection method, from 1.7 with the sucrose diet to 3.7 (P<0.001). Proliferation measured by the incorporation of ^{3}H thymidine gave similar results. Using the comet assay to assess genotoxic damage, it was found that DMH-induced DNA damage in the colon was decreased significantly (P<0.05) by CrystaLean compared with sucrose.

Bacteriological profile was also altered by CrystaLean in both UK and Italian flora associated HFA rats. Mean $\log_{10}$ counts of lactobacilli and bifidobacteria were increased (P<0.001) by CrystaLean in each case and enterobacteria counts were decreased by CrystaLean in both UK (P<0.01) and Italian (P<0.001) flora rats.

4 CONCLUSIONS

Metabolic events associated with the gut microflora and potentially involved in colon carcinogenesis (β-glucuronidase activity, ammonia concentration and 7-OHIQ formation) were highest in the sucrose-fed group and lowest in the CrystaLean (resistant starch)-fed group. The CrystaLean diet was also found to modify, in a potentially beneficial manner, the bacteriological composition of the colonic microflora in rats colonised with either UK or Italian flora.

Proliferation, assessed by both the microdissection method and by incorporation of ^{3}H thymidine, revealed that crypt cell proliferation was highest in the rats fed CrystaLean and statistical analysis showed a significant positive correlation with cecal n-butyric acid and colonic bile acid concentrations.

In the Comet assay, DNA damage induced by *in vivo* treatment with DMH (15mg/kg) was greatest in the rats fed sucrose and lowest in the CrystaLean-fed group.

It would appear therefore that the type of carbohydrate fed can have significant effects on bacterial metabolic processes, microfloral composition and mucosal changes in the colon relevant to carcinogenesis and that despite stimulating colonic mucosal proliferation, CrytaLean exhibited potentially beneficial effects towards colonic neoplasia.

5 ACKNOWLEDGEMENT

The authors wish to acknowledge the financial support for this research programme by EU (AIR2-CT94-0933 and AIR-1-CT94-7122) and MAFF.

6 REFERENCES

1. C.J Rumney and I.R. Rowland, *Crit. Rev. Fd. Sci. Nutr.*, 1992, **31**, 299.
2. G. Caderni, F. Bianchini, P. Dolara and D. Kriebel, *Nutr. Cancer*, 1991, **15**, 33.
3. C.J. Rumney, I.R. Rowland and I.K. O'Neill, *Nutr. Cancer*, 1993, **19**, 67.
4. B.L. Pool-Zobel, B. Bertram, M. Knoll, R. Lambertz, C. Neudecker, U. Schillinger, P. Schmezer and W.H. Holzapfel, *Nutr. Cancer*, 1993, **20**, 271.
5. G.W. Snedecor and W.G. Cochran, 'Statistical Methods 6th Edition', Ames: Iowa State University Press.

IN VITRO FERMENTATION OF ENZYMATICALLY DIGESTED RYE BREAD: PREPARATION OF THE FECAL INOCULUM

A.M. Aura, A. von Wright, M. Alander, M. Fabritius, T. Suortti and K. Poutanen

VTT Biotechnology and Food Research
Espoo
Finland

1 INTRODUCTION

Cereal dietary fibre (DF) includes β-glucans and pentosans such as arabinoxylans. Traditional Finnish wholemeal rye bread is a good source of arabinoxylans, but β-glucans are also an important component of the rye grain cell wall material. The effects of processing on the cereal cell wall matrix may be reflected in the digestibility and fermentability of DF. In order to decrease the number of variables in the fermentation and to study the changes in the DF during bacterial action, a model with pooled and frozen fecal inoculum is introduced. Thus, the differences in the fermentation are due only to the fibre material: its chemical composition and process-induced changes. Moreover, the reproducibility of fermentations can be improved by using the same inoculum for a longer period of time.[1] The *in vitro* method introduced in this paper is a simple batch application of the continuous model of Molly *et al.*[2]

It is vital for the fermentation studies to use as representative a human colonic flora as possible, and to use it in an adequate amount. It is important to detect the losses of bacterial species in the preparation of inoculum and during fermentation. Two fecal inocula were compared with respect to their bacterial species as well as for their ability to produce short-chain fatty acids (SCFA; acetate, propionate and butyrate) in the fermentation of wholemeal rye bread as the sole substrate. The effect of freezing on the inoculum was also studied.

2 MATERIALS AND METHODS

In the first experiment, feces from five healthy volunteers were homogenized and diluted with nitrogenised water under a nitrogen stream with the Stomacher to a final concentration of 90mg dry wt/ml and frozen in Eppendorf vials at -70°C. In the second experiment, 250g feces from six healthy volunteers, 225g peptone-phosphate buffer (BPW, Oxoid, pH 6.7) containing sodium thioglycolate (0.5g/L) and cysteine-hydrochloride (0.5g/L), and 25g glycerol were homogenized in an anaerobic chamber using the Waring Blender as a homogeniser. The final concentration of the second inoculum was 85mg dry wt/ml slurry.

Commercial sourdough wholemeal rye bread was digested with alimentary enzymes according to Aura *et al.*[3] Rye bread fibre (300mg) was fermented with fecal inoculum (1) 0.09mg dry wt/batch or (2) 8.5mg dry wt/batch in fibre-free medium according to Molly *et al.*[2] All the procedures with feces prior to fermentation were carried out in an anaerobic chamber. Fermentation was performed at 37°C at 150rpm for 24 hours and at the end of fermentation the samples were centrifuged (6000rpm, 11 minutes) and SCFA were analysed as free acids.

In the first experiment anaerobic and aerobic bacteria were cultivated on plate count agar; enterobacteria were detected on violet red-bile-glucose agar and *Clostridia* on Tryptose-sulfite-cycloserine agar. In the second experiment anaerobic and aerobic bacteria were detected on Brain-heart infusion agar.

The SCFA of samples (2ml) with an internal standard (heptanoic acid) were extracted twice with 2.5ml diethyl ether and analysed with a gas chromatograph (Micromat HRGC 420) using an FI detector and helium as the carrier gas. The SCFA of the blanks without fibre were subtracted from the results.

3 RESULTS AND DISCUSSION

In the first experiment counts of anaerobic bacteria decreased in the freezing procedure. Counts of both the total anaerobes and aerobes were 4.2×10^8 colonic forming units (cfu)/gram dry weight of feces after freezing. Division to Eppendorf vials made the inoculum susceptible to oxygen and facultative anaerobes were the only species surviving this step (Figure 1). The bacterial population was significantly changed during fermentation of rye fibre as substrate. The concentrations of anaerobic and aerobic bacteria remained the same as in the frozen inoculum. The concentration of *Enterobacteria* was increased, whereas *Clostridia* and *Bacteroides* were totally absent after fermentation.

In the first experiment (0.09mg dry weight fecal inoculum/batch) 410mmol acetate and 23mmol propionate/g dry weight feces were produced. The relatively high amount of SCFAs was due to the dilution of the inoculum. The SCFA profile and also the lack of butyrate indicated unbalanced anaerobic fermentation of rye fibre. The most significant result of bacterial action detected in the fermentation with the frozen first inoculum was the decrease of MW of β-glucans of rye fibre (Figure 2).

In the second experiment the numbers of anaerobic and aerobic bacteria were 3.2×10^9 and 3.9×10^8 cfu/g dry weight feces, respectively. After fermentation of rye fibre with fresh inoculum the growth of anaerobic bacteria was promoted: the amounts of anaerobic and aerobic bacteria were 6.4×10^{11} and 7.4×10^8 cfu/g dry weight feces, respectively, indicating better anaerobiosis in the fecal fermentation than in the first experiment. When fresh inoculum (8.5mg dry wt/batch) was used in the second experiment 26mmol acetate, 9mmol propionate and over 50mmol butyrate were produced/g dry weight feces. Production of butyrate indicated better anaerobiosis and was in agreement with previous studies on rye fibre fermentation.[4]

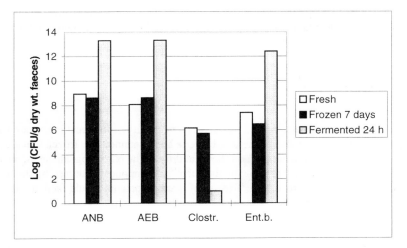

Figure 1 *Amounts of different bacteria in the first fecal inoculum before and after freezing and after fermentation as Log(cfu/g dry wt feces). (ANB: anaerobic bacterial count on PCA; AEB: aerobic bacterial count on PCA; Clostr: Clostridia on TSCA; Ent.b: Enterobacteria on VRBGA.)*

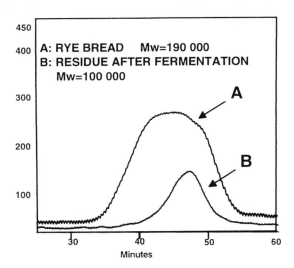

Figure 2 *Weight average molecular weight GPC chromatograms of β-glucans before and after fermentation of rye bread with frozen fecal inoculum (0.09 mg dry wt/batch). Detection with post-column Calcofluor staining and a fluorescence detector according to Jaskari et al.[5]*

In conclusion, the preparation of the inoculum requires strict anaerobic conditions throughout the procedure. Rye fibre seemed to promote the growth of butyrate producers and other anaerobic bacteria.

4 FURTHER STUDIES

Further experiments are being carried out to verify the quality of the fecal inoculum. Can the balance between anaerobic and aerobic bacteria be maintained in a frozen inoculum? What are the effects of fermentation on the fecal flora? The physiological behaviour of consumer foods is influenced not only by the chemical components but also by the food structure created during processing. The use of frozen inoculum would facilitate the *in vitro* studies of these process effects.

5 REFERENCES

1. A. Stevenson, C.J. Buchanan and M.A. Eastwood, *J. Sci. Food Agric.*, 1997, **73**, 93.
2. K. Molly, M.V. Woestyne, I.D. Smet and W. Verstraete, *Microb. Ecol. Health Dis.*, 1994, **7**, 191.
3. A.-M. Aura, M. Koskinen, K. Autio, H. Härkönen and K. Poutanen, Poster, Plant Polysaccharide Symposium, INRA, Nantes, 17.-19.7. 1996.
4. W. Feldheim and E. Wisker. 'International Rye Symposium: Technology and Products, VTT Symposium Series 161, Espoo, 1995, 93.
5. J. Jaskari, K. Henriksson, A. Nieminen, T. Suortti, H. Salovaara and K. Poutanen, *Cereal Chem.*, 1995, **72**, 6

CHARACTERISTICS OF *LACTOBACILLUS CASEI* SHIROTA, A PROBIOTIC BACTERIUM

C. Shortt

Yakult UK
Acton, London
W3 7XS

1 INTRODUCTION

Lactobacillus casei Shirota (LcS) is a lactic acid bacterium which was isolated in 1930 by Dr Minoru Shirota, a medical microbiologist at Kyoto University, Japan. In 1935, Dr Shirota developed the unique LcS fermented milk drink, Yakult, which is available world-wide today. A freeze-dried form of the LcS strain is also available as a licensed product called Biolactis Powder. Studies demonstrating that LcS has properties characteristic of probiotic lactobacilli are reviewed.

Probiotic microorganisms can only be utilized effectively if they fulfil a number of scientific criteria. These include general (origin, safety, tolerance to gastrointestinal barriers), technological (production and processing characteristics) and functional criteria.[1]

2 GENERAL CHARACTERISTICS

Strains isolated from the human intestinal tract are generally recommended for probiotic use. LcS is of human origin and has the following morphological and biological properties:

(i)   gram positive, straight single or chained rods
(ii)   facultative anaerobe
(iii)   0.6-0.7μm in width
(iv)   1.5-5.0 μm in length
(v)   grows between 15 and 41 °C.

The safety profile of probiotic strains used in the food supply is critical. Extensive *in vitro* and *in vivo* studies have been conducted using LcS and products derived from the fermentation of LcS. The strain has been well-accepted and tolerated in all the human volunteer trials performed to-date. In addition, data are available to support the lack of pathogenicity of the strain in animal toxicological studies.

Acid and bile tolerance are essential properties for any strain selected to have effects in the intestinal tract.[1] Studies have shown that LcS tolerates synthetic gastric juice well (Figure 1) and exhibits tolerance to synthetic bile. In contrast, the classical yoghurt strains, *Streptococcus thermophilus, L. bulgaricus* and *L. jugurti* are less tolerant to synthetic gastric juice.[2]

3 TECHNOLOGICAL CHARACTERISTICS

From a technological perspective, it is essential that the stability of the required characteristics and the viability of a probiotic strain are maintained during production and storage. LcS survives well in processed milks. It is known that optimal fermentation processing can enhance strain characteristics. *In vitro* studies have shown that the tolerance of LcS to gastrointestinal barriers is enhanced by the six-day fermentation process. The survival of LcS in the stomach of the dynamic *in vitro* TNO gastrointestinal model, was higher in the fermented milk product than in an unfermented milk suspension which had been innoculated with an overnight culture of the strain.[1]

4 FUNCTIONAL CHARACTERISTICS

Functional characteristics, such as, modulation of the intestinal flora, antagonism to pathogens, adherence, stimulation of the immune response and modulation of specific intestinal metabolic activities are important in the selection of probiotic strains. LcS

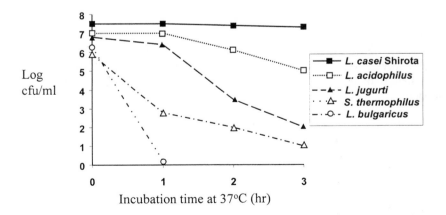

Figure 1 *Survival of selected lactic acid bacteria in synthetic gastric juice (pH 3.0)*
        (adapted from reference 2)

exhibits weak *in vitro* adherence to Caco-2 cell lines compared with several other Lactobacillus species (Table 1).[3] The *in vitro* adhesion results suggest that LcS binds non-specifically and indicate that approximately 3 x 10^6 bacteria were bound in the case of LcS compared to 1 x 10^7 bacteria in the case of the most adherent strains.[3] However, the relevance of such findings to *in vivo* adherence and efficacy is unclear at present.

5 HUMAN VOLUNTEER STUDY

LcS has been shown to influence the nature and activity of the intestinal flora in several studies. A study, evaluating the effects of LcS fermented milk consumption, on healthy Dutch volunteers corroborates results from previous studies carried out in Japan.[4,5] Twenty healthy male volunteers, aged 40 to 65 years, participated in the eight-week study. The protocol consisted of a two-week baseline period, a four-week test period and a two-week follow-up period. The test group received 300ml of LcS fermented milk (10^9 colony forming units (cfu)/ml) per day and the control group received a similar amount of un-fermented milk (<10 cfu/ml). The volunteers stayed in a metabolic suite during the last three days of weeks 2, 4, 6 and 8 and samples of feces, urine and blood were collected for microbiological, biochemical and immunological analyses.

Recovery of administered LcS was high (10^7 cfu/g wet weight feces) in the group that consumed the LcS fermented milk (Figure 2). Fecal LcS levels returned to baseline levels (10^2/g wet feces) within 14 days following cessation of consumption of the fermented milk. Results demonstrate the ability of LcS to survive passage through the intestinal tract where it persists transiently.[5] In the test group total fecal lactobacilli counts increased, bifidobacteria counts significantly increased and the number of fecal clostridia decreased. Activities (per 10^{10} bacteria) of two bacterial enzymes (β-glucuronidase and β-glucosidase) that may be involved in the conversion of procarcinogens to proximal carcinogens decreased.[5]

Table 1 *Ranking of Lactobacillus Strains Based on Adherence to Intestinal Caco-2 Cell Cultures (adapted from reference 3)*

Bacterial strain	Rank	Approx. no of bacteria bound
L. casei 744	1	
L. acidophilus LA1	2	
L. rhamnosus LC-705	3	
L. GG ATCC 53103	4	1 x 10^7
L. rhamnosus ATCC 7469	5	
L. rhamnosus (human isolate)	6	
L. planatarum ATCC 8014	7	
L. casei	8	
L. casei Imunitass	9	
L. casei 01	10	
L. casei Shirota	11	3 x 10^6
L. casei var. rhamnosus	12	

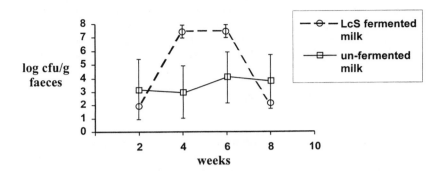

Figure 2 *Effect of ingestion of L. casei Shirota fermented milk and un-fermented milk on mean L. casei Shirota counts in the feces of healthy Dutch volunteers* (adapted from reference 5)

No significant differences in general health parameters (body weight, blood pressure), blood chemistry (serum cholesterol, albumin, globulins, glutamic oxaloacetic transaminase (GOT) and glutamic pyruvate transaminase (GPT)) and immune parameters including natural killer (NK) cell activity, phagocytosis, cytokine production and humoral parameters were observed (Table 2).

Table 2 *Effect of ingestion of L. casei Shirota fermented milk and unfermented milk on selected immunological parameters in healthy Dutch volunteers (adapted from reference 5)*

Immunological paramater	Control group				Test group			
	wk2	wk4	wk6	wk8	wk2	wk4	wk6	wk8
T helper (CD4+),%	45±8	46±9	48±9	47±9	44±8	45±6	47±6	45±7
T suppressor/cytotoxic (CD8+),%	34±5	33±6	32±6	33±6	36±9	34±9	33±8	34±8
NK (CD16 & 56+),%	21±8	21±10	18±8	19±10	22±10	19±7	18±8	21±9
IFNγ, x10pg/ml	176±99	138±71	193±123	193±106	117±48	113±62	108±94	99±53
ILI-β, x10pg/ml	84±26	84±38	92±24	109±41	84±23	73±48	84±35	106±36
IL-2, x10pg/ml	60±30	58±33	50±21	63±40	40±28	48±31	46±22	49±29
Phagocytosis neutrophils,%	57±14	56±16	54±15	52±12	55±6	56±11	51±12	47±8
Oxidative burst neutrophils, %	19±9	24±6	22±8	16±7	20±8	21±7	19±4	15±9

This paper presents data demonstrating that LcS has properties characteristic of probiotic lactobacilli and that consumption of LcS fermented milk influences the composition and metabolism of the intestinal microflora but has no effect on the immune system of healthy males. Recent studies have also shown that probiotic strains, including LcS, have been efficacious in various therapeutic settings.[6]

6 REFERENCES

1. J.H.J. Huis in't Veld and C. Shortt, Selection criteria for probiotic micro-organisms. In: Gut Flora and Health - Past, Present and Future. Eds Leeds, A. & Rowland, I. International Congress and Symposium Series 219. The Royal Society of Medicine Press Ltd, London, 1996, 27-36.
2. Y. Kobayashi *et al.* Studies on biological characteristics of lactobacillus 11. *Japanese Journal of Bacteriology* 1994, **29**, 691-697.
3. E.Lehto and S.Salminen, Adhesion of twelve different Lactobacillus strains to Caco-2 cell cultures. *Nutrition Today* 1996, **31**, 49s-50s.
4. R.Tanaka, The effects of the ingestion of fermented milk with Lactobacillus casei Shirota on the gastrointestinal microbial ecology in healthy volunteers. In: Gut Flora and Health - Past, Present and Future. Eds Leeds, A. & Rowland, I. International Congress and Symposium Series 219. The Royal Society of Medicine Press Ltd, London, 1996, 37-45.
5. S. Spanhaak and R.Havenaar, *TNO Report*, 1993, **V 92**, 691-697.
6. S.Salminen *et al.* Clinical uses of probiotics for stabilising the gut mucosal barrier: Successful strains and future challenges. *FEMS Microbiology Reviews*, 1996, **70**: 347-358.

II Evidence for Health Benefits

FUNCTIONAL FOODS - CAUTIONARY NOTES

A.E. Bender

2 Willow Vale
Fetcham Leatherhead
Surrey KT22 9TE

1 INTRODUCTION

Terms such as 'major new product opportunity', 'a new dimension', 'a new global added-value market', 'a major opportunity for the dairy industry' indicate the opinions of company marketing divisions. However, there are two main reasons for advising caution in the potential promotion of functional foods.

The first is a general caution arising from public suspicion of processed foods, stimulated largely through scaremongering. Consumers are wary of all new developments ranging chronologically from irradiation to genetic manipulation, with food poisoning thrown in as an additional confusing factor.

Consumers are likely to view with scepticism claims for new discoveries of the special health virtues of something as ordinary as food and so further distrust the food industry. Whether or not such beliefs are correct, public perception is that (a) 'natural' products are safer, better and preferred to those that have been 'messed about with,' and (b) while 'natural' foods are good for you they are so well-established and ordinary that they could not really have any special (and new) health-promoting properties.

The second is the difficulty of supplying evidence for claims. Scientists have spent half a century in trying to find the 'cause' of coronary heart disease (CHD), including dietary factors—what resources are needed to 'prove' the virtues of the large number of proposed functional foods?

There is epidemiological evidence that various patterns of food consumption are associated with decreased risk of cancer and CHD. This has been extrapolated to certain nutrients, vitamin E, carotene etc. Good biochemical mechanisms have been suggested and measurements of plasma levels of these nutrients correlate well with the incidence of cancer and CHD. Yet intervention studies with many of these nutrients have been disappointing. We do not know that these nutrients are the protective or risk factors; they may be markers for the intake of other compounds. For example, in the USA there is a highly significant correlation between the intake of saturated fat and plasma total carotenoids; an even better correlation with lycopene specifically. Hamburgers soaked in tomato ketchup probably explains this.

The content of beneficial factors in some foods can be increased by genetic manipulation—presumably converting them into functional foods. A growing number of these substances such as aromatic isothiocyanates, indoles and organic sulphur compounds have been shown to block chemical carcinogenesis—but also shown to be genotoxic or tumorigenic.[1] There would appear to be risk as well as benefit. As these authors state "the message of epidemiology is that vegetables and fruits are overwhelmingly beneficial in their effects but the prospect of manipulating the composition of commercial varieties inevitably raises issues of safety" (Table 1).

Extrapolation from controlled trials that show benefits of specified ingredients may not apply to whole diets and varied lifestyles. Any one food, other than a staple, cannot be expected to overcome the effects of a diet that is composed of thousands of ingredients and varies from day to day in both types and amounts. Can all consumers expect the same benefits? Have trials continued long enough to guarantee long-term safety? Only now is the consumer becoming aware of the long latent period of, for example, BSE (bovine spongiform encephalothapy), although scientists have long known that the effects of certain carcinogens become evident only after 20 years or more.

2 OVERDOSAGE

2.1 Dietary Fibre

Dietary fibre has been shown to be beneficial in a range of disorders such as cancer of the colon, atherosclerosis, diverticulosis, hypertension and obesity, so why not add soluble fibre to all our beverages? Would the intake reach harmful levels?

Despite evidence of intakes of 150g dietary fibre/day in parts of Africa, US authorities recommend not more than 35g (total) fibre/day. The effects of fibre include alterations of the intestinal flora (with or without the presence of various types of 'functional' lactobacilli) and alteration of the response of intestinal cells and gut hormones and changes of the structure of the intestinal villi and goblet cell numbers.

Excess fibre can give rise to intestinal gas production and discomfort, reduced absorption of zinc, iron and calcium and even to intestinal obstruction.

In Japan, currently leading in the development of fibre, some 40% of functional foods on the market supply dietary fibre and it is expected that changes in plant genetics and processing together with modifications of existing forms of fibre will lead to further development. Some of the types of fibre and products available are shown in Table 3.

2.2 Iron

Iron is a toxic substance that is under physiological control but some 10% of the population of the USA carry genes that cause excessive absorption of iron: 1 in 500 males in the USA suffer from iron deficiency; 2 in 500 suffer from haemochromatosis; approximately 10% have one gene for enhanced iron absorption, 1 in 250 have two genes. It has been recommended that before taking iron supplements or additional vitamin C (which enhances iron absorption) subjects should be tested for their ferritin levels.[3] A product called Hemace in Japan incorporated into candy contains purified haem and is claimed to be absorbed 30 times as efficiently as other forms of iron. Free

Table 1 *Naturally Occurring Compounds in Plant Foods that are Carcinogenic in Experimental Animals*

	Major sources
Bergaptin	celery, parsley, parsnips
Benzil acetate	basil, honey, jasmine tea, anise, apple, carrot, celery, cherry, dill, endive, grapes, lettuce, marjoram, pear, plum, rosemary, tarragon, thyme, tomato
Cathetol	coffee beans
Coumestrol (phyto-oestrogen)	alfalfa, soybean sprouts
Cycasin	cycad nuts
Estragole	anise, basil, fennel, tarragon
Genistin (phyto-oestrogen)	soybean, clover
P-hydrazinobenzoates	mushrooms
Ipomeanol, ipemearone	sweet potatoes
Isoflavones	soybean
Lubminin	potatoes
Luponic acid	hops
Methyl benzyl alcohol	cocoa
Methylpsoralens	celery, parsley, parsnip
Mirestrol (phyto-oestrogen)	legumes
Oestrone	palm kernels
Phaseolin	green beans
Pisatin	peas
Pyrrazolidone alkaloids	coltsfoot, comfrey, Crotolaria spp., Heliotropum spp., Senecio spp.
Rishitin	potatoes
Safrole	mace, nutmeg, pepper (black), sassafras oil
Sesamol	sesame oil
Sinigrin (allylisothiocyanate)	brussels sprouts, cabbage, cauliflower, collard greens, horseradish, mustard (brown)
Santhotoxin	celery, parsley, parsnips
Zearalone (phyto-oestrogen)	Fusarium roseum (growing on maize)

Source: Bender and Bender, 1987.[2]

iron ions not bound to transferrin, ferritin or haemosiderin undergo non-enzymic reactions leading to the formation of oxygen radicals.

2.3 Amino Acids

We can already buy, at prices vastly greater than the same amounts in bread, every amino acid as supplements. Should we market functional foods enriched with amino acids when we learn that while arginine inhibits tumor growth methionine enhances tumour growth?[4]

2.4   Antioxidants

It is recognised that increased intakes of various antioxidants may be beneficial but which ones and how much? In low doses they function as antioxidants: in high doses as pro-oxidants, especially in the presence of iron and other minerals eg copper.

It has been reported that very high doses of vitamin C (5g/day) which are poorly absorbed can result in diarrhoea and can throw AIDS patients into shock, because it is superimposed on existing diarrhoea.

3 GENETIC NUTRITION

Genetic make-up renders individuals susceptible to a range of diseases which might not ensue until triggered by some environmental factor. For example, there appears to be a genetic inability to process cholesterol which may not show up clinically unless the subject has a regular diet rich in fats.

Similarly the genetic susceptability to cancer may not develop into malignancy until the individual is exposed to carcinogens, smoking or the presence or absence of certain foods.

We already know that dietary changes can bring benefit in some genetic diseases—and many are more likely to follow.

3.1   Circulatory Problems

With regard to low density lipoprotein (LDL) and heart problems it has been shown that subjects with small dense particles of LDL in their blood compared with large and more buoyant particles, have a substantially lower LDL cholesterol and apoprotein B. If such subjects were identified they would benefit from a low fat diet. In general, diets could well be tailored to suit the biochemical systems and lifestyles of individuals.

This application of 'functional medicine' depends on knowing, as early as possible in life, the genetic susceptabilities of the individual. This is far removed from wholesale dosage of the public with a range of foods that may be beneficial to some and possibly not to others—or even harmful to some.

It has been shown that women with the apo E4 gene respond to polyunsaturated fatty acids by reducing their blood levels of beneficial HDL cholesterol due to increased absorption of dietary cholesterol (and so are recommended to increase their intake of monounsaturated fatty acids).

Table 2 *Sources of Dietary Fibre*

Bran from oats, wheat, rice; peas, bagasse,pectin Gums from locust bean, guar, xanthan, gellan; psyllium,aloe vera Pullulan (Aureobasidium pullulans) Curdlan (Alicaligenes faecalis) Dextran (from sucrose) Fibercel (from yeast)
Products available in Japan Fibemini PF 21 (sports drink with collagen) Meiji Seika cereal (plus fibre) Fibre-mix 391 Fibi - fibre-rich soft drink Cheese enriched with fibre

It is reported that 10% of the USA population may carry a genetic predisposition for CHD because of poor metabolism of homocysteine. This can be modified by increasing the intake of vitamin B_6, folate and vitamin B_{12} to stimulate the conversion of homocysteine to methionine.

Whether oat or wheat bran will or will not reduce blood cholesterol level is under genetic control. Those whose cholesterol decreases in response to dietary fibre are those with two particular gene variants, the EZ allele and a change in the base pair at a specific site of LOL receptor. The intakes of energy, and amounts of fat, carbohydrate, fibre and protein will vary with age, weight, metabolism and activity level—all except age have a genetic component.

The Rochester Minnesota Family Heart study[5] illustrates the interaction between genes and diet (Table 3). Both the genetic differences, (Finns with much higher frequency of apo E4 gene) and diet (Finns with 24% energy from saturated fat and Japanese with 3%), are reflected in the enormous difference in male mortality from CHD.

Simopoulos *et al.*[6] suggested that people should compile a family health tree in an attempt to discover whether they, as individuals are likely to carry beneficial or harmful genes—such as problems with iron absorption, blood pressure, certain cancers—to ascertain whether they need to abstain from certain food ingredients or can ingest them without likely harm, whether they could overcome the effects of sodium with calcium or potassium-rich foods, extra magnesium, fish, etc.

So a functional food beneficial for some may be harmful to others.

4 CONCLUSION - PUBLIC PERCEPTION

Apart from scientific problems there is the question of public perception—the

Table 3 *Male CHD v E4 Gene and Cholesterol Levels*

Country	Mortality[1]	% Energy from saturated fat	E4 Gene[2]	Cholesterol
Finland	700	24	2x	2x
Minnesota	350	16	1x	1.5x
Japan	130	3	1x	1x

[1]per 100,000 population
[2]frequency of apo E4 gene relative to Japan.

ultimate arbiter.

It has taken accidents and tragedies to put food in the headlines—though too many have been distorted. So, we have intense (media) interest in food, nutrition and safety, even if the public lags behind the media (and the scientists).

However, along the way questions are asked that cannot be readily answered, eg how can you be sure that current *safe* levels of intakes of, for example, additives/residues are really safe over a lifetime? How do you know that a combination of additives/contaminants/adulterants, especially after partial breakdown in cooking, is still safe, and is it safe for everyone? Add the combined effects of pollution and all that is implied and can you still ensure safety? Of course, no one can ensure complete safety, particularly since people continue to cross the road and drive cars. So the public (or is it only the media?) asks questions that are becoming more and more difficult to answer.

Not long ago we might well have dismissed or ignored such questions but this is no longer possible. How do we know that ingesting substance X for 50 years might not shorten life, or affect resistance to disease or alter the quality of life? We do not know, but the very questions illustrate concern and hence emphasise caution.

Finally, when there is good evidence how can we persuade consumers that the product and the claims are genuine? The health food industry abounds with claims that the product has been scientifically tested, as proven by thousands of satisfied customers, or the claims for the product in question are cunningly inserted into orthodox text in a way that is misleading.

Whatever the current loosening of legislation (deregulation) and the unwillingness of Ministers to control functional foods, events and the public have overtaken them.

5 REFERENCES

1. I.T.Johnson, G.Williamson, S.R.R.Musk, *Nutr.Res.Rev.* 1994, 7, 175-204.
2. D.A.Bender, A.E.Bender, "Nutrition - a reference handbook". Oxford University Press. p 481. 1997.
3. A.P.Simopoulos, V.Herbert, B.Jacobson, "Genetic Nutrition" Macmillan, New York. p 161. 1993.

4. W.J.Visek, *Cancer Res,* 1992, **52**, 20825-20845.
5. A.P.Simopoulos *et al ibid.,* p 7.
6. A.P.Simopoulos *et al ibid.,* p 44.

ANTIOXIDANT NUTRIENTS

J.J. Strain[1] and I.F.F. Benzie[2]

[1]Human Nutrition Research Group, University of Ulster, Coleraine BT52 1SA
[2]Department of Health Sciences, The Hong Kong Polytechnic University Kowloon, Hong Kong

1 INTRODUCTION

There is a substantial amount of data which indicates a major role for diet in the development of a range of inflammatory disorders including coronary heart disease (CHD) and cancer. The precise components, however, of the diet which cause harm or confer benefits are much less clear. Nevertheless, one of the most consistent research findings is that people who consume higher amounts of fruit and vegetables have lower risk of CHD and various cancers, especially lung and stomach but also cancer of the colon and breast.[1,2]

The strong and consistent associations in epidemiological studies, however, do not prove causality. High fruit and vegetable consumption might simply appear protective because people who eat more of these are likely to be more physically active, smoke fewer cigarettes, drink less alcohol and be more medically and nutritionally aware than those whose diet is poor in fresh fruit and vegetables.[3] Moreover, it is very difficult to disentangle lifestyle factors from possible dietary factors, especially given that the measurement of habitual diet in free living individuals is fraught with difficulty and the importance of exposure to a causal factor may extend back to childhood or the fetal environment.

If a diet rich in fruit and vegetables offers protection against the common chronic diseases of adulthood (and the weight of evidence from observational studies after allowing for confounding factors, strongly suggests that they do),[1,4] then there are various hypotheses which could be considered to explain this. The most prominent of these is the antioxidant hypothesis which holds that antioxidants offer protection from disease by minimising or preventing the damage caused by oxidants.[5] Although fruit and vegetables are rich sources of antioxidants, these protective substances are also found in other foods and there is the obvious potential for the application of antioxidant nutrients (and non-nutrients) to functional food products.

2 OXIDANTS AND OXIDATIVE DAMAGE

Potentially damaging reactive oxygen species (ROS) are produced during normal oxidative metabolic processes, during inflammation and after ingestion of certain drugs and exposure to pollutants. Cigarette smoke, abnormal oxygenation and radiation along with lipid oxidation products in food may also contribute to the ROS flux or oxidant stress within the body. Sources of ROS, therefore, can be divided into either physiological ('normal') or pathological ('abnormal'). The former includes ROS production which is purposeful, directed, useful, but potentially damaging if excessive, prolonged or uncontrolled eg production of ROS (specifically superoxide and hypochlorous acid) by white cells to fight infection or the release of nitric oxide for maintenance of normal vascular tone. Normal physiological oxidative processes of energy production can also lead to accidental, potentially harmful production of ROS.

It is probable that various ROS cause oxidative damage *in vivo*. The problem with identifying those ROS found *in vivo* is that many are so short lived that they are only detectable as 'foot-prints' ie they leave behind oxidized products which indicate their past existence. These ROS are capable of oxidizing all major biomolecules and this may lead to widespread disruption of cellular metabolism, carcinogenesis, atherogenesis, auto-immune changes and cell death. Although oxidative changes have been implicated in a wide range of diseases including CHD, cancer, diabetes, stroke and arthritis, these diseases are inflammatory in nature and the oxidized 'footprints' associated with the disease may arise simply as a result of more active white cell production of ROS. Oxidative changes, therefore, might be secondary to the disease rather than causative.

3 ANTIOXIDANTS

Important biological defence systems have evolved to limit inappropriate exposure to ROS.[5,6] Intracellular, extracellular, lipophilic and aqueous antioxidant mechanisms are found throughout the body and work in concert to prevent generation of ROS, to destroy or inactivate ROS which are formed or to terminate chains of ROS-initiated peroxidation. Many endogenous and exogenous constituents of biological tissues or fluids show antioxidant (reducing) activity in *in vitro* studies and at least some of these may have important antioxidant actions *in vivo*. When antioxidant defences are inadequate owing to poor nutrition or increased exposure to ROS, oxidant stress may occur and oxidative damage may increase. Dietary intake of antioxidants is important in replenishing and augmenting these defences and, thereby preventing or at least minimising such damage.

Human tissues are generally rich in a variety of scavenging and chain-breaking antioxidants. These include antioxidants which must be supplied pre-formed in the diet such as ascorbic acid (vitamin C), α-tocopherol (vitamin E) and possibly carotenoids (including β-carotene), as well as endogenously produced antioxidants such as bilirubin, uric acid, ubiquinol and glutathione. Dietary trace elements are vital components of enzymic antioxidant defence mechanisms such as caeruloplasmin (copper), superoxide dismutase (copper, manganese), glutathione peroxidase (selenium) and also for metalloenzymes (zinc) needed for DNA repair. The major mineral, magnesium, and the B-vitamin, riboflavin are co-factors in the production of reduced glutathione. In addition transition metals such as iron and copper, which can degrade pre-existing peroxides with the subsequent formation of ROS, are kept out of the peroxidation equation by being tightly bound to, or incorporated into, specific proteins.

The antioxidant hypothesis has been supported to date mainly by *in vitro* experimental evidence. *In vivo* epidemiological data, however, are accumulating and are now quite extensive for the two apparently most important non-enzymic antioxidants, vitamins C and E and also for β-carotene.[7-10] Although most research has focussed on these, increasing attention has been given to other carotenoids (lycopene, lutein and zeaxanthin) and also to flavonoids, a large group of phytochemicals found in fruit and vegetables and which comprise of flavonols, flavones, catechins, flavones and anthocyanidins.[11,12] It should be noted, however, that many of these nutrients and phytochemicals may also give protection from disease via mechanisms other than antioxidant defence.

Observational population studies which have investigated the links between some of these antioxidants and cancer have found that β-carotene seems to offer much more protection than the antioxidant vitamins. In order to determine, however, whether a specific antioxidant component (or components) in fruit and vegetables really is protective, human intervention studies, using either single components or mixtures, must be conducted. The results of three major preventive studies, which have investigated the protective effect of pharmacological (much larger than normal dietary) doses of β-carotene supplements, have found that, contrary to expectations, β-carotene is not protective and may actually be harmful, especially with respect to incidence of lung cancer among smokers.[13-15]

Findings from intervention studies on CHD, which have investigated the protective effects of vitamin E—apparently the antioxidant most likely to confer protection from observational data—have also been somewhat disappointing. Small doses of vitamin E appear to offer little protection.[15] One study[16] which used pharmacological doses found a reduction in non-fatal myocardial infarction but not in fatal myocardial infarction or total mortality. Therefore, at least until the results of other intervention studies are published, there appears to be no substitute for a healthy diet rich in fruit and vegetables.

4 RICH DIETARY SOURCES OF ANTIOXIDANTS

The current dietary recommendation to increase fruit and vegetable consumption is one which is widely perceived as health-promoting and the consistent epidemiological links worldwide between high fruit and vegetable consumption and a greater life expectancy warrant more emphasis given to this particular dietary recommendation. Fruit and vegetables are rich sources of the antioxidants, vitamin C, vitamin E, various carotenoids, flavonoids, isoflavonoids, organo-sulphur compounds, copper, manganese and magnesium and may also contribute to pools of endogenously produced antioxidants such as ubiquinol (Table 1). Fruit and vegetables, however, are not the only dietary source of antioxidants and other rich sources of vitamin E include nuts and seeds, wholegrain breakfast cereals, wholemeal bread, eggs, margarine, vegetable oils, and dairy products. The carotenoids are also not restricted to fruit and vegetables and dairy

Table 1 *Plant-based Sources of Antioxidants*

(i)	Vitamin C (ascorbic acid) - blackcurrants, green peppers, guavas, gourds, greens, strawberries, kiwi fruits, citrus fruits, paw-paws, brussel sprouts, new potatoes
(ii)	Vitamin E - sweet potatoes, spinach, broccoli, pulses, kale, tomatoes, asparagus, herbs
(iii)	Carotenes - carrots, sweet potatoes, herbs, pumpkins, spinach, greens, kale, cantaloupes, chicory, squashes, red peppers, mangoes, apricots
(iv)	Lycopene - guavas, pink grapefruits, tomatoes
(v)	Lutein and zeaxanthin - kale, spinach, herbs, greens, celery, scallions, leeks
(vi)	Flavonoids - onions, strawberries, apples, citrus fruits, greens, broad beans, peanuts, grapes, tea
(vii)	Isoflavones - pulses, especially soya bean and linseed products
(viii)	Organo-sulphur compounds - allium vegetables: garlic, onions, chives, leeks
(ix)	Ubiquinol - beans, garlic, spinach
(x)	Copper - pulses, mushrooms, olives, gourds, avocados, lychees, blackberries, blackcurrants, kiwi fruits, grapes, mangoes, guavas, bananas, raspberries, plums, asparagus, potatoes
(xi)	Manganese - beetroot, blackberries, pineapples, pulses, spinach, greens, bananas, raspberries

Adapted from reference 5.

products are an important source of carotenoids in some Western diets.

Cereal based foods, especially bread, are important contributors to dietary selenium, although levels of selenium and, to some extent, copper in foods are dependent on the trace element content of the soil in which foods have been grown. Consistently rich sources of selenium, however, are seafoods and liver. Seafoods, particularly shellfish, and offal are also rich sources of copper as are nuts and seeds, wholemeal bread and wholegrain breakfast cereals and chocolate. Meat (especially liver), milk and dairy products will supply adequate pre-formed vitamin A (another possible antioxidant *in vivo*) and zinc. Milk and dairy products are also a major dietary source of another antioxidant, riboflavin. It is obvious, therefore, that a balanced diet is important in the maintenance of adequate antioxidant defences.

Functional food products enriched with antioxidants could be especially helpful to individuals whose dietary intakes of antioxidants might be sub-optimal because of insufficient dietary variety or for other reasons. While high intake of products enriched with some antioxidants, such as vitamin E, may not constitute a health hazard, care should be taken that consumers are not exposed to excessive intakes of some other antioxidants, such as other fat soluble vitamins or the trace elements[17] through high consumption of such functional food products.

5 CONCLUSIONS

The observed health benefits of eating fruit and vegetables are remarkably consistent across population studies but it is currently unclear which of the many biologically active constituents (including antioxidants) of fruit and vegetables confer protection against chronic diseases. The evidence for a protective role for individual antioxidant nutrients (mainly vitamins E and C and β-carotene) which have been tested to date in intervention studies is disappointing. Although this evidence militates against the intake of antioxidant supplements to increase antioxidant defence, many people appear to resort to such means to augment antioxidant intakes. This suggests that there is a market niche for foods enriched with antioxidant nutrients or with other antioxidant phytochemicals.

6 REFERENCES

1. M.W. Gilman, *B.M.J.*, 1996, **313**, 765.
2. M.G.L. Hertog, H.B. Bueno-de-Mesquita, A.M. Fehily, P.M. Sweetnam, P.C. Elwood and K. Kromhout, *Cancer Epidemiol. Biomarkers Prev.*, **5**, 673.
3. M. Nestle, *Nutr. Rev.*, 1996, **54**, 255.
4. A.R. Ness and J.W. Powles, *Chem. Ind.*, 1996, 792.
5. J.J. Strain and I.F.F. Benzie, 'Encyclopedia of Human Nutrition', M. Sadler, B. Cabellero and J.J. Strain (ed), Academic Press, London (in press).
6. B. Halliwell, *Ann. Rev. Nutr.*, 1996, **16**, 33.
7. L.H. Kushi, A.R. Fobsom, R.J. Prineas, P.J. Mink, Y. Wu and R.M. Bostick, *N. Eng. J. Med.*, 1996, **334**, 1156.
8. N.R. Sahyoun, P.F. Jacques and R.M. Russell, *Am. J. Epidemiol.*, 1996, **144**, 501.
9. G. van Poppel, *Eur. J. Clin. Nutr.*, 1996 **50** (suppl 3), 557.
10. K. Nyyssönen, M.T. Parviainen, R. Salonen, J. Tuomilehto and J.T. Salonen, *B.M.J.*, 1997, **314**, 634.
11. N.C. Cook and S. Samman, *J. Nutr. Biochem.*, 1996, 7, 66.
12. P. Knekt, R. Järvinen, A. Reunanen, J. Maatela, *B.M.J.*, 1996, **312**, 478.
13. The Alpha-tocopherol, Beta-carotene Cancer Prevention Study Group, *N. Engl. J. Med.*, 1994, **330**, 1029.
14. G.S. Omenn, G.E. Goodman, M.D. Thornquist, J. Balmes, M.R. Cullen, A. Glass, J.P.Keogh, F.L. Meyskens, B. Valanis, J.H. Williams, S. Barnhart and S. Hammar, *N. Eng. J. Med.*, 1996, **334**, 1150.
15. C.H. Hennekens, J.E. Buring, J.E. Manson, M. Stampfer, B. Rosner, N.R. Cook, C. Belanger, F. LaMotte, J.M. Gaziano, P.M. Ridker, W. Willett and R. Peto, *N. Eng.*

J. Med., 1996, **334**, 1145.

16.N.G. Stephens, A. Parsons, P.M. Schofield, F. Kelly, K. Cheeseman, M.J. Mitchinson and M.J. Brown, *Lancet*, 1996, **347**, 781.

17.D.P. Richardson, *Proc. Nutr. Soc.* (in press).

ANTIOXIDANT PROPERTIES OF THE ISOFLAVONE-PHYTOESTROGEN FUNCTIONAL INGREDIENTS IN SOYA PRODUCTS

H. Wiseman, J.D. O'Reilly, P. Lim, A.P. Garnett, W-C Huang and T.A.B. Sanders

Nutrition, Food and Health Research Centre
King's College London
Campden Hill Road
London W8 7AH

1 INTRODUCTION

Dietary phytoestrogens may help to protect against atherosclerosis and cancer. The most abundant isoflavone phytoestrogens are genistein and daidzein, which are found in soya beans (*Glycine max*), and soya products (mostly in their glycoside conjugate forms). The traditional soya foods (fermented and nonfermented), widely consumed by populations in the Far East, are produced from soya beans.[1] Traditional fermented soya foods include soya sauce, tempeh, natto, miso and fermented tofu and soya milk products. Soya milk is the aqueous extract derived from whole soya beans. The nonfermented soya foods include whole-fat soya flour, whole dry beans, soya milk and the soya milk product tofu. In soya beans, textured vegetable protein (TVP) and tofu (soya bean curd) there are high levels of the conjugated isoflavones daidzin and genistin, whereas in the fermented soya bean product miso, most of the isoflavones are present in their unconjugated forms.[2]

Phytoestrogens may exert health protective effects by modulation of oestrogen metabolism and through their other biological properties, which include antioxidant activity. Following ingestion daidzin and genistin are usually considered to be hydrolysed in the large intestine by the action of bacterial glycosidases to release genistein and daidzein. Daidzein can be metabolised by bacterial enzymes to form equol (oestrogenic) or *O*-desmethylangolensin (non-oestrogenic), whereas genistein is metabolised to the hormonally inert *p*-ethylphenol.[3] Interindividual variation in ability to metabolise daidzein to equol could thus influence the potential health protective effects of soya isoflavones.[3] Furthermore, a pilot study of isoflavone absorption from TVP suggests that absorption of intact isoflavone glycones may be possible.[3]

We have investigated some of the antioxidant properties of soya isoflavones in several model systems. We found that the aglycone isoflavones genistein and daidzein were generally weak antioxidants in model membrane systems including the liposomal membrane model. The glycoside conjugates of these isoflavones, genistin and daidzin displayed no antioxidant action in these systems. However, equol, the oestrogenic isoflavan metabolite of daidzein, was a potent antioxidant in these systems. In contrast, the compounds were generally more effective at protecting isolated human low density lipoprotein (LDL) against oxidative damage than preventing liposomal membrane lipid peroxidation. Furthermore, a protective effect against LDL fatty acid peroxidation was demonstrated for genistin.

2 ANTIOXIDANT PROPERTIES OF SOYA ISOFLAVONES

Liposomes are used extensively as a model membrane system for studying the influence of dietary components and drugs on membrane lipid peroxidation *in vitro*. Liposomes are artificial lipid structures, and are made by shaking or sonicating phospholipids in aqueous suspension. In this study ox-brain phospholipid was suspended in phosphate buffered saline (140mM NaCl, 2.7mM KCl, 16mM Na_2HPO_4, 2.9mM KH_2PO_4) at pH 7.4 at a final concentration of 10mg/ml, followed by sonication.[4] The resulting milky suspension of liposomes was allowed to stand in at 4°C for one hour prior to use. Liposomal membrane peroxidation in the presence of Fe(III) and ascorbate was measured by the formation of thiobarbituric acid reactive substances (TBARS) as described previously.[4]

A wide range of dietary components have been reported to protect human LDL against oxidative modification.[5] This may be of importance because oxidative damage to LDL (particularly to the apoprotein B molecule) may be an important stage in the development of atherosclerosis. Lipid peroxidation begins in the polyunsaturated fatty acids of the phopholipids on the surface of LDL and then propagates to core lipids resulting in modification of the phospholipids, polyunsaturated fatty acids, cholesterol and the apolipoprotein B molecule. In most studies on the action of dietary components on oxidative damage to LDL, human LDL is stimulated to undergo lipid peroxidation by the addition of Cu(II) ions.[5]

In this study LDL was isolated from human plasma in a Beckman ultracentrifuge using a SW40Ti (swing bucket) rotor. Fresh plasma (no older than one day old) was obtained from a blood bank and sucrose was added to a final concentration of 10%.[6] The plasma was then aliquoted into plastic tubes and frozen at -70°C until required (though for no longer than three months). The plasma was centrifuged for 18 hours at 100,000g at 22°C. LDL was dialysed for 18 hours against 10mM PBS pH 7.4 to remove EDTA. The concentration of protein was determined by the modified Lowry method. The LDL was filtered through a Sartorius 0.2µm filter unit after dialysis and stored for up to four days. LDL lipid peroxidation in the presence of Cu(II) was measured by the formation of thiobarbituric acid reactive substances as described previously. LDL (0.1mg/ml) in a final volume of 0.5ml of 10mM PBS pH 7.4 was incubated at 37°C with $CuSO_4$ (at a final concentration of 5µM) for two hours. Test compounds were added in ethanol to give a final concentration of ethanol in the reaction mixture of 1% (v/v) and this concentration of ethanol was always used in controls. The reaction was terminated by the addition of 10µl of 10mM BHT dissolved in ethanol and the extent of lipid peroxidation was determined by the thiobarbituric acid (TBA) test.

The results (Table 1) show that the isoflavones genistein, daidzein and their conjugates were relatively poor inhibitors of membrane lipid peroxidation in the liposomal system. Indeed, only genistein displayed significant activity. This is in marked contrast to the potent antioxidant ability displayed by the daidzein metabolite equol, which was similarly effective to 17 β-oestradiol and 4-hydroxytamoxifen. In contrast, the isoflavones, including the genistein conjugate genistin, were considerably more potent inhibitors of LDL fatty acid peroxidation than of liposomal lipid peroxidation.

The isolated isoflavones, related compounds, extracts of soya milk and soya milk itself were all tested for interaction between their membrane antioxidant action and that of 2.5μM 17 β-oestradiol (Table 2). The percentage differences between the theoretical and observed values for inhibition of liposomal membrane lipid peroxidation indicates the extent of antagonism that occurred. The highest levels of antagonism were observed with the isoflavone-rich extract of both sweetened and unsweetened soya milk and also with isolated daidzein, and to a lesser extent genistein (Garnett and Wiseman, unpublished results). Antagonistic membrane antioxidant interactions with 17 β-oestradiol were also observed for both tamoxifen and 4-hydroxytamoxifen. Isoflavone-rich extracts of soya beans from different countries of origin, were able to inhibit liposomal lipid peroxidation by 10-20%, and block the antioxidant effect of 30μM 17 β-oestradiol by as much as 33% (Lim and Wiseman, unpublished results).

The antagonistic interactions observed between the soya isoflavones and 17 β-oestradiol suggest that the antioxidant action of each is being inhibited by the other. The molecular basis of the observed antagonisms may be associated with stereo-specific or regio-specific interactions of pairs of the antagonistic molecules. The blocking of the interaction of these 'dimers' with the membrane lipid bilayers may then be possible. These results suggest that soya isoflavones could, if present in the body at suitable levels, interact and block the antioxidant action of the endogenous hormone 17 β-oestradiol, perhaps in a similar manner to their antagonism of 17 β-oestradiol at the oestrogen receptor.

In further experiments both sweetened and unsweetened soya milk and the isoflavone-rich extracts prepared from them, were found to have an inhibitory effect on the membrane antioxidant activity of both black tea and green tea extracts, in terms of their ability to inhibit liposomal lipid peroxidation (Huang and Wiseman, unpublished results) and we are currently investigating the underlying mechanisms involved.

Table 1 *Protection by Isoflavones and Related Compounds against Liposomal Membrane Lipid Peroxidation and against Oxidative Damage to Human LDL*

Compound	Liposomal system TBARS IC$_{50}$ (μM)	LDL system TBARS IC$_{50}$ (μM)
Genistein	50	8
Genistin	NR	26
Daidzein	NR	13
Daidzin	NR	NR
Equol	5	3
17 β-Oestradiol	9	2
Tamoxifen	30	15
4-Hydroxy-tamoxifen	8	1

Values are deduced from graphs (not shown) in which each point represents the mean±SEM of at least three separate assays. NR - not reacted

Table 2 *Antioxidant Interaction between Isoflavones, Related Compounds and 17 β-Oestradiol*

Compound/extract added with 2.5µM 17 β-oestradiol	Theoretical result % inhibition of liposomal TBARS	Observed result % inhibition of liposomal TBARS	% Difference Indicative of extent of antagonism
Genistein (25µM)	47±4	30±3	36
Daidzein (30µM)	34±6	18±1	47
Equol (2.5µM)	41±4	25±3	40
Tamoxifen (25µM)	49±6	28±4	43
4-Hydroxy-tamoxifen (2.5µM)	39±4	25±1	36
Sweetened soya milk extract (10µl)	53±8	26±4	50
Unsweetened soya milk extract (10µl)	62±9	33±6	46
Soya milk (10µl)	40±6	32±6	2

Values represent the mean±SEM of at least three separate assays

3 SOYA ISOFLAVONES AND PROTECTION AGAINST HORMONALLY RELATED CANCER

The consumption of soya products is associated with a lower risk of prostate and breast cancer in some populations. Tofu but not miso consumption was associated with a lower risk of fatal prostatic cancer in men of Japanese ancestry in Hawaii followed for 20 years.[7,8] The latency period for prostate cancer appears to be lengthened in men of Japanese origin in Hawaii, who have a low mortality from prostate cancer. The incidence of *in situ* prostate cancer in autopsy studies is similar, however, to that of men in Western countries. It has been postulated that the consumption of soya isoflavones by these men may be responsible for this long latency period. A high intake of miso soup has been associated with a reduced risk of breast cancer in Japanese women.[7,8] A case-control study in Singapore also found the consumption of soya products to be associated with a decreased risk of breast cancer.[7,8]

Carcinogenesis involves three different stages, initiation, promotion and progression. Free radicals may play a role in human cancer development. Indeed, free radicals have been shown to possess many characteristics of carcinogens.[9] Mutagenesis by free radicals could contribute to the initiation of cancer, in addition to being important in the promotion and progression phases.[9] The reported antioxidant properties of soya isoflavones could thus contribute to their proposed anticancer action. However, the isoflavones may also exert their protective effects by antagonising the action of sex hormones and by interfering with cellular signalling mechanisms involving tyrosine kinase.[10] Phytoestrogen isoflavones appear to block the growth promoting action of oestrogen (and androgens) on cancer cells. Genistein and daidzein appear to have some structural homology to the female sex hormone oestrogen. In order for oestrogen to stimulate the growth of breast cancer cells it first needs to bind to its cell surface receptor, the oestrogen receptor. Isoflavones can also bind to this receptor because of their structural mimicry but do not produce the same stimulatory effects on cell division. Thus soya isoflavones may effectively act as antioestrogens, perhaps in a similar way to the drug tamoxifen.[5,10] Tamoxifen is widely used in the treatment of breast cancer and currently being assessed for this disease with cardioprotective benefits.[5,10]

Dietary isoflavones are excreted into breast milk. This suggests that in mothers who routinely consume soya products, breastfed infants are likely to be exposed to high levels of soya isoflavones, apparently without any adverse effects. Furthermore, isoflavone exposure through breastfeeding occurs at a critical developmental period and may make an important contribution to any protective effects against cancer.[11] Indeed, this may be a more important factor than adult dietary exposure to isoflavones, in the observation of lower cancer rates in populations in the Far East.

In 17 of 26 animal studies of experimental carcinogenesis, utilising diets containing soya or isolated soya isoflavones, protective effects were reported. The risk of cancer (incidence, latency or tumour number) was greatly reduced. Furthermore, no studies reported that soya intake increased tumour development.[7,10] In addition, in rats treated neonatally with 7,12-dimethylbenz[a]anthracene, genistein was found to delay the appearance of mammary tumours. This was found to be associated with increased cell differentiation in the mammary tissue. In *in vitro* cancer models, genistein inhibited the proliferation of human cancer cells in culture. Genistein is a specific inhibitor of tyrosine kinases, which are involved in a signalling cascade that ultimately leads to cell division.[7,10] Cancer cells spread by inducing the growth of new blood vessels by a process called angiogenesis. Genistein appears to be a potent inhibitor of angiogenesis. This antiangiogenic activity of genistein could be useful in the treatment of cancer.[7,10]

4 CARDIOPROTECTIVE EFFECTS

A lower incidence of coronary heart disease has been reported in populations such as the Japanese and US Seventh Day Adventist vegetarians,[12] where soya products are routinely consumed. However, the diets in these populations also tend to be lower in saturated fat and cholesterol. Nevertheless, protection against atherosclerosis may be another potential health benefit of phytoestrogens. Studies in primates have shown that soya isoflavones prevent atherosclerosis and have beneficial effects on lipoprotein metabolism.[13] A meta-analysis of human dietary trials has shown that soya protein can reduce cholesterol and LDL and raise high density lipoprotein (HDL).[14] This may be mediated via the oestrogenic effects of phytoestrogens on the liver where they probably induce the expression of LDL receptors. Indeed, removal of phytoestrogens from soya reduces its cholesterol lowering properties in primates.[13] Preventing the oxidation of LDL is viewed as important in the oxidative hypothesis of atherosclerosis. Oxidised LDL-cholesterol is taken up by macrophages within the sub-endothelial cells compartment of the arterial wall to form foam cells. Excessive foam cell formation leads to fatty streak formation and apoptosis of the foam cells is believed to lead to fibrous plaque formation. HDL is believed to play an important role *in vivo* in protecting against oxidation of LDL. Our results show that the oestrogenic isoflavones can protect LDL against oxidative modification *in vitro*, (the isoflavan metabolite equol was particularly potent). We are currently investigating whether LDL oxidation is inhibited in subjects who have been fed soya isoflavones.

5 CONCLUSIONS

The antioxidant action of the isoflavones present in functional foods containing soya may contribute to the potential health benefits of these foods by protecting against the oxidative damage implicated in many disease states.[9] However, it seems likely that their overall mode of action is more complex and also involves endocrine disruption and modulation of cell signalling mechanisms. Phytoestrogens have been reported to cause infertility in some animals and concerns have been raised over their consumption by human infants especially as hypospadias and other feminization effects have become more common. The influence of a high phytoestrogen intake on reproductive hormone concentrations in infants, and whether any changes could cause disrupt the imprinting of subsequent sexual behaviour, is an issue that needs investigating. The oestrogens widely used in HRT and the oral contraceptive pill are known to adversely influence haemostatic variables, such as levels of blood clotting factors. This could lead to increased susceptibility to thrombosis. The effects of soya phytoestrogens on these factors is currently under investigation.

6 ACKNOWLEDGEMENT

This project is funded by MAFF (UK)

7 REFERENCES

1. P.Golbitz, *J. Nutr.* 1995, **125**, 570S.
2. L. Coward, N.C. Barnes, K.D.R. Setchell, and S. Barnes, *J. Agric. Food Chem.*1993, **41**, 1961.
3. E.A. Bowey, I.R. Rowland, H. Adlercreutz, T.A.B. Sanders and H. Wiseman, RSC/BNF Functional Foods'97 conf. proc. (in press).
4. H. Wiseman, *Biochem. Pharmacol.*1994, **47**, 493.
5. H. Wiseman, *J. Nutr. Biochem.* 1996, **7**, 2.
6. J. O'Reilly, L. Pollard, D. Leake, T.A.B. Sanders and H. Wiseman, *Proc. Nut. Soc.* 1997 (in press).
7. H. Adlercreutz, B.R. Goldin,. S.L. Gorbach, K.A.V. Hockerstedt, S. Watanabe, E.K. Hamalainen, M.H. Markkanen, T.H. Makela, K.T. Wahala, T.A. Hase, and T.Fotsis, *J. Nutr.* 1995, **125**, 757S.
8. A. Cassidy, *Proc. Nutr. Soc.* 1996, **55**, 399.
9. H. Wiseman and B. Halliwell, *Biochem. J.* 1996, **313**, 17.
10. H. Wiseman *Biochem Soc Trans* 1996, **24**, 795.
11. A.A. Franke and L.J. Custer, *Am. J. Clin. Nutr.* 1997 (in press).

12. P.K. Mills, W.L. Beeson, R.L. Phillips and G.E. Fraser, *Am. J. Clin. Nutr.* 1994, **59** 1136S.

13. M.S. Anthony, T.B. Clarkson, C.L. Hughes Jr, T.M. Morgan and G.L. Burke, *J. Nutr.* 1996, **126** ,43.

14. J.W. Anderson, B.M. Johnstone, M.E. Cook-Newell, *New Engl. J. Med.* 1995 **333,** 276.

A TOXICOLOGIST'S VIEW ON DIETARY CHEMOPREVENTION

H. Verhagen

TNO Nutrition and Food Research Institute
PO Box 360
3700 AJ Zeist
The Netherlands

1 INTRODUCTION

The paradox of nutrition is that it is both essential to support life and is thought to play a major role in the causation of some cancers. All nutrients are necessary to sustain life. Nutrients can be subdivided into macronutrients and micronutrients. Non-nutritive dietary constituents can be subdivided into compounds with adverse potential ('toxicity') and compounds with beneficial potential ('beneficity').

In this paper the current interest in bioactive dietary constituents will be discussed in the light of the established science of toxicology. A series of caveats is presented be taken into account when claiming that a food (ingredient) has chemopreventive potential.

2 TOXICOLOGY COMPARES DOSE AND EFFECT

In toxicology, health risk assessment is in essence based on only two aspects: dose (amount per weight or volume) and effect (response, including underlying mechanisms). All substances are toxic. Each compound has a threshold dose at and below which no effect will occur—the 'no-observed-adverse-effect-level' (NOAEL). The NOAEL is commonly generated in studies with experimental animals: several doses are tested in animals ranging from clearly toxic doses via a 'lowest-observed-adverse-effect-level' (LOAEL) down to the NOAEL. It is assumed that the NOAEL established in animals is also the NOAEL in humans. In order to overcome possible intra- and interspecies differences, a potentially safe level for human exposure (eg the acceptable daily intake, ADI) is calculated by dividing the NOAEL by a safety factor (SF): $ADI = NOAEL/SF$ (Figure 1).

This NOAEL/SF approach is applicable to any compound, except to non-stochastically acting non-genotoxic carcinogens and perhaps allergens. Also in pharmacology and nutrition 'dose' is the factor that discriminates between the presence or absence of an effect. In the case of medicines or putative beneficial dietary constituents, a desired effect may be called 'lowest effect level' (LEL). In the case of nutrients 'dietary reference values' are set: 'lower reference nutrient intake' (LRNI), 'estimated average requirement' (EAR), 'reference nutrient intake' (RNI), and 'safe intake'. In Table 1 a survey is given of the dose and effect requirements for nutrients, non-nutrients and medicines.

3 TOXICOLOGY AND DIETARY CHEMOPREVENTION

Toxicology provides us with the lines of evidence necessary to establish a true chemopreventive effect in humans. As such there are several caveats to take into account, all of which are based on the concept of toxicology:
Health risk assessment = f (dose, effect).

3.1 Caveats Related to Dose

3.1.1 Caveat 1: The Threshold Concept. Chemoprevention is necessarily a non-stochastical event; *ie* it cannot be assumed that in theory one molecule can prevent carcinogenicity. Also for the establishment of beneficial effect levels the threshold principle applies. Thus, there will be a LEL for a chemopreventive effect to become manifest, *ie* exposure to putative beneficial substances below the LEL remains necessarily without effect. This is far from new—even for medicines a high enough dose is needed to have the desired effect (eg to cure a disease).

3.1.2 Caveat 2: Beware of Toxicity! For putative chemopreventive substances the toxicological (NOAEL, LOAEL) and beneficial dose levels (LEL) should be considered together in a single evaluation. A beneficial effect is only valuable in the

Table 1 *Putative Effects and Dose Requirements for nutrients, non-nutrients and Medicines*

Effect:	Necessity (for sustaining life)	Beneficity (towards human health)	Toxicity (towards human health)
dose requirement:	LRNI ≤ dose ≤ RNI	LEL ≤ dose < NOAEL	LOAEL ≤ dose
macro-nutrient	yes	yes	yes
micro-nutrient	yes	yes	yes
non-nutrient with beneficial potential	no	yes	yes
non-nutrient/toxicant	no	no	yes
medicine	sometimes (specific to a disease)	yes (by definition)	yes

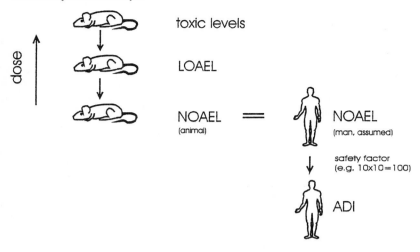

Figure 1 *Extrapolation of animal toxicity data towards safe levels for humans*

absence of toxicity: the LEL should be well below the safe human dose. Hence the beneficial effects should be evident at (much) lower dose levels than where toxicity is expected.

For nutrients RNI and NOAEL levels may be very close and a proper SF may not be feasible; rather a 'margin of safety' would apply (Figure 2). In contrast, with medicines, toxic side effects may be unavoidable. In such cases the necessity of therapy outweighs the concominant toxicity. In case of dietary chemopreventive agents it is not acceptable to have toxicity at beneficial dose levels.

3.1.3 Caveat 3: The Matrix Considered. Dietary constituents are not consumed as pure substances. Rather they are inevitably parts of our daily food. In toxicology it is recognized that man is simultaneously exposed to a huge number of chemicals. There is uncertainty as to how the combined toxicity of these chemicals should be assessed and how combined toxicity should be taken into account in setting standards for individual compounds. This equally well applies to chemoprevention. Indeed, in general it is unlikely that single compounds may be consumed in sufficient quantity to elicit the desired effects. However, a combination of beneficial substances in a matrix ('the food') may result in beneficial effects in humans by means of additivity, synergism or potentiation, although antagonising effects may also occur. Moreover, by spreading the beneficial effects over the combined action of a number of substances undesirable toxic effects may be avoided.

3.2 Caveats Related to Effect

3.2.1 Caveat 4: Assessment of Chemopreventive Potential. Genotoxicity of a compound is generally tested in a tiered approach. First, short-term *in vitro* tests with prokaryotic or eukaryotic cell systems are performed, followed by short-term *in vivo* tests in experimental animals. Depending on the results of the short-term genotoxicity tests, a long-term *in vivo* study in experimental animals may be performed, in which the carcinogenic potential of a compound is established by life-time exposure of experimental animals to various dose levels of the test compound up to some level of

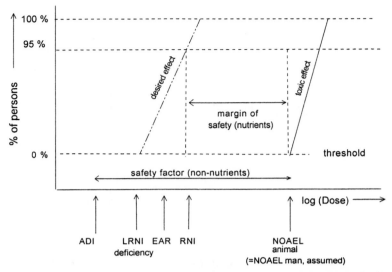

Figure 2 *Theoretical and simplified dose-effect relationships for desired effects (eg nutrients, medicines, dietary chemopreventive agents) and for toxic effects. In general, the curves are not parallel, they may cross over, or the curve for toxicity may be at lower dose levels than for beneficial effects. The NOAEL is divided by a SF to obtain an ADI for man. Below the LEL there is no beneficial effect whatsoever (eg for chemopreventive agents, medicines), whereas for nutrients below the LRNI there will be a deficiency. The band between the RNI and the NOAEL constitutes a 'margin of safety' rather than a SF*

toxicity. Occasionally also human data are available from epidemiology and/or biomarker studies.

All these experimental test systems for genotoxicity *in vitro* and *in vivo* and carcinogenicity *in vivo* can equally well be applied to determine the chemopreventive potential of compounds. In addition, there are some very rapid chemico-analytical methods available such as for the assessment of antioxidant potential. An antimutagenic response *in vitro* should be verified *in vivo*, bearing in mind that *in vitro* test systems have limitations of high target cell concentrations and direct exposure of target cells. An antigenotoxic effect in short-term *in vivo* studies can be overruled if there are no indications for anticarcinogenicity in long-term animal studies or in human studies.

3.2.2 Caveat 5: The Underlying Mechanism. Denominating a compound as a carcinogen only on the basis of functionality is rather a poor toxicological approach. Evidence should be provided on the mechanism that is involved in the carcinogenic activity. As such, a discrimination is made between genotoxic and non-genotoxic carcinogens which is of utmost importance for health risk assessment. The same holds true for denominating an anticarcinogen. If the underlying mechanism is not known a claim of anticarcinogenic potential has a weak basis. The mechanisms of chemopreventive agents are multiple. The multistage nature of carcinogenesis raises the possibility for intervention at each stage of the process as well as many modes of action for chemopreventive agents, such as:

i)   prevention of formation/uptake of carcinogens

ii)   scavenging effect on the (activated) carcinogens
iii)   shielding of nucleophilic sites in DNA
iv)   inhibition of DNA/carcinogen complex
v)   modifying effect on the activities of xenobiotic-metabolizing enzymes,
vi)   modifying effect on the activities of other enzymes
vii)   antioxidative activity
viii) other ....

3.2.3 Caveat 6: (Anti)carcinogens are not Always (Anti)mutagens and Vice Versa. In the early 1970s toxicologists thought that carcinogens could be found by performing short/term genotoxicity tests *in vitro* and *in vivo*. Indeed, initially there was a steadily growing overlap between these two categories of compounds, especially when the use of liver homogenate, 'S9', was introduced to *in vitro* assays. In later years the overlap decreased again; carcinogens were sometimes, but not always, mutagens and *vice versa*. One of the main reasons for this is that nowadays the rodent carcinogenicity assays are overly sensitive because of the necessity to test at a 'maximum tolerated dose', thereby rendering almost every second compound a carcinogen. In this way many a 'carcinogen' is a non-genotoxic carcinogen (and thus in fact a non-carcinogen). Similarly, anticarcinogens may not always be antimutagens and *vice versa*.

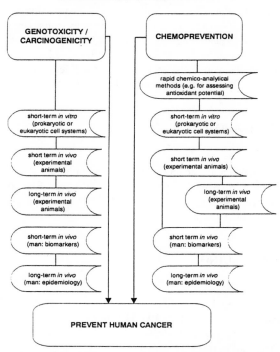

Figure 3 *Tiered approach for assessing carcinogenicity and anticarcinogenicity*

3.2.4 Caveat 7: Weight of the Evidence. In toxicology conclusions are seldom drawn on the basis of just one study. A whole series of data is necessary to make a complete judgement of the toxicological status of a compound. For synthetic industrial agents, pesticides, food additives and medicines, a large number of reports conducted according to officially recognized (eg OECD) guidelines and performed under eg Good Laboratory Practice are obligatory prior to obtaining permission for the market introduction of a compound. The compilation of the dossier can take years or even a decade. Upon delivery of the whole set of data a final judgement can be made taking into account all data from regular toxicity studies as well as information on the mechanism of toxicity. The most critical endpoint is taken as the starting point for performing a health risk assessment. By using a similar argument the chemopreventive potential of conpounds should not be claimed on the basis of just one or a few studies.

4 THE PROOF OF THE PUDDING .... IS IN THE EATING

Given the caveats for chemoprevention described above one might wonder whether it is feasible to identify these desired effects in humans. However, there are indications to underpin the feasibility of dietary chemoprevention in man. These may be epidemiological studies based on dietary questionnaires and/or experimental studies using biomarkers. Epidemiological studies have indicated that several dietary habits are associated with a decreased risk of cancer, such as fruits, vegetables, and fibre. In fact such epidemiological findings have triggered the onset of experimental chemoprevention studies *in vitro* and *in vivo*. Also experimental biomarker studies in human volunteers have indicated that it is indeed feasible in humans to have a potential beneficial effect in the absence of adverse effects.

5 ACKNOWLEDGEMENTS

The following scientists are acknowledged for stimulating discusssions: V.J. Feron, G. van Poppel, C.J.M. Rompelberg, M. Strube, P.J. van Bladeren (TNO Nutrition and Food Research Institute, Zeist, The Netherlands), G. Williamson (IFR, Norwich, UK), M. Smith (Unilever, Colworth, UK).

6 FURTHER READING

1. B. N. Ames & L.S. Gold, *Proc. Natl. Acad. Sci. USA.* 1990, **87**, 7772-7776.
2. L.R. Ferguson, *Mutat. Res.* **307**, 395-410, 1994.
3. V.J.Feron et al., *Toxicology Letters*, 1995, **105**: 415-427.
4. C.D. Klaassen, 'Casarett and Doull's Toxicology', 3rd Ed. (C.D. Klaassen *et al.* eds.), Macmillan Publ. Co., Chapter 2, 1986.
5. P. Mason, 'Handbook of dietary supplements', Blackwell Science, 1995.

6. NRC, 'Carcinogens and anticarcinogens in the human diet', Natl. Acad. Press, 1996.

7. H. Verhagen and G. van Poppel, 'Proceedings of the FIE '96 Functional Foods Seminar', Paris, France, 13 Nov 96 (in press), 1997.

8. H. Verhagen *et al.*, 'Food Chemical Risk Analysis', David Tennant (Ed.), Chapman and Hall (Publ.). (in press), 1997

9. L.W. Wattenberg, *Cancer Res.* 1992, **(Suppl.) 52**, 2085s-2091s

THE EFFECT OF GUAR-CONTAINING WHEAT BREAD ON THE GLYCAEMIC RESPONSE IN ILEOSTOMATES

S.J. Hurley,[1] J. Tomlin,[1] P.R. Ellis,[2] S.B. Ross- Murphy[2] and L.M. Morgan[3]

[1] Sheffield University, Department of Surgical & Anaesthetic Sciences- K Floor, The Royal Hallamshire Hospital, Sheffield S10 2JF
[2] Biopolymers Group, King's College London, London W8 7AH
[3] Surrey University, Department of Biological Sciences, Guildford, Surrey

1 INTRODUCTION

Guar galactomannan is a water soluble non-starch polysaccharide (s-NSP). As is characteristic of s-NSP it is not digested by specific enzymes in the upper gastrointestinal tract and therefore its presence is likely to exert a physiological effect.

The physiological effects of guar are well documented in the literature. Guar's effects on carbohydrate metabolism have been studied for many years.[1-3] Following the consumption of guar, which has been added to glucose drinks or carbohydrate-rich meals, postprandial rises in glucose and insulin are delayed.

s-NSP, such as guar, form 'entangled networks' when hydrated, resulting in viscous solutions *in vitro*.[4] Guar is thought to improve glycaemic control by virtue of its ability to increase viscosity within the gastrointestinal tract. This has been shown to occur in pigs, but has not been demonstrated in humans thus far.[5]

The mode of delivery and formulation of guar gum is of paramount importance in obtaining the optimum physiological effects. Guar is currently available as an antidiabetic drug, which is added to a pre-prandial drink or sprinkled onto foods. However the clinical efficacy of such products is poor. This is thought to be due to variations in hydration rates. It has been demonstrated that intimate mixing of guar gum into food products improves its effects on glycaemic control.[6,7] The molecular weight and concentration of guar galactomannan affects the rate of hydration which is the rate limiting step in obtaining optimal viscosity.[8]

1.1 Aims

The aim of the present study was to investigate guar's effect on intestinal viscosity and carbohydrate metabolism in humans. Ileostomy patients were used in the study, in order to access intestinal contents for viscosity measurements.

This is the first attempt in humans to relate the physiological effects of guar to intestinal viscosity and therefore establish a link between its physico-chemical properties and mechanism of action.

2 METHODS

2.1 Experimental Design

Seven ileostomy patients with established stomas took part in the study, which was approved by the local Ethics Committee. Four females and three males took part; the age range was 33-70 years with a median of 56 years.

After an overnight fast recruits came to the Department of Surgery and Anaesthetics, where they were fitted with a Venflon cannula, kept patent with Heplok (10 units of heparin per ml). Basal blood samples were taken for fasting glucose, glucose-dependent insulinotrophic polypeptide (GIP), glucagon-like peptide 1,7-36 amide (GLP-1) and insulin measurements.

Stoma bags were then emptied and the recruits were given a bread meal and 150ml water, to provide 75g available carbohydrate. On one occasion the meal was a control test bread and on the other occasion the bread contained approximately 5% guar galactomannan, incorporated into the bread as a flour (M90 grade, Meyall Chemicals AG; molecular weight 1×10^6). Both meals were nutritionally balanced, containing equal proportions of fats, proteins and carbohydrates.

Postprandial venous blood samples were taken at the following time points: 30, 45, 60, 90, 120, 150 and 180 minutes. Following the meal the recruits' stoma bags were checked regularly for contents and samples were collected as frequently as possible and frozen immediately for viscosity analysis at a later date.

2.2 Blood Sample Assays

Glucose analysis was carried out using the glucose oxidase method (YSI instruments). GIP and GLP-1 were analysed at Surrey University using an in-house double antibody radioimmunoassay technique.

3 RESULTS

Following consumption of the guar-containing bread meal, venous glucose, GIP and GLP-1 measurements were reduced.

Figure 1 shows mean (n=7) postprandial changes in plasma glucose, in relation to fasting values, after consumption of either a control or guar test meal. Venous glucose was reduced following the guar meal as compared to the control, at every time point sampled, except 180 minutes. The difference was significant at 90 minutes postprandially (*p=<0.05 Wilcoxon signed- rank test).

GIP and GLP-1 results are only available for five of the recruits. Figure 2 shows plasma GIP in relation to fasting values after each test meal. GIP is reduced at each time point sampled, following guar consumption with significant differences seen at 90 and 180 minutes postprandially (*p<0.05 Wilcoxon signed-rank test).

Figure 3 shows plasma GLP-1 values, after each test meal. Venous GLP-1 is lower after guar at all sampling points except 15 and 180 minutes, with significant differences seen at 30 and 60 minutes postprandially (ADP<0.05, Wilcoxon signed-rank test).

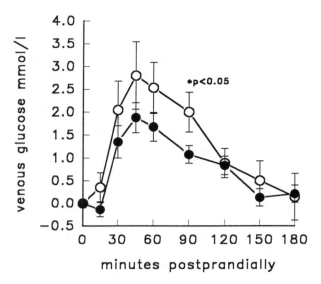

Figure 1 *Postprandial glucose in relation to fasting values following a guar or control test meal (n=7). Hollow circles represent the control meal; filled circles represent the guar meal. Standard error bars are shown*

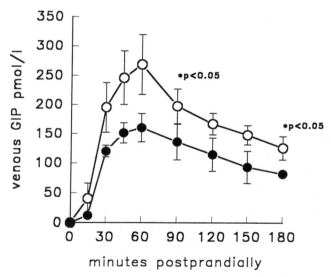

Figure 2 *Postprandial GIP in relation to fasting values, following a guar or control test meal (n=5). Hollow circles represent the control meal; filled circles represent the guar meal. Standard error bars are shown*

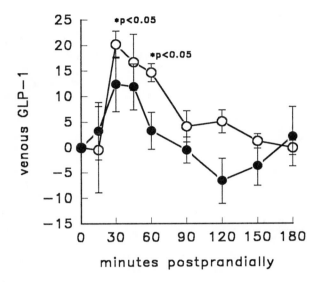

Figure 3 *Postprandial GLP-1 in relation to fasting values, after a guar or control test meal (n=5). Hollow circles represent the control meal; filled circles the guar meal. Standard error bars are shown*

Guar consumption significantly altered the rheological properties of gastrointestinal contents collected from the recruits' stomas (the results are not shown).

5 DISCUSSION

The gastrointestinal hormones GIP and GLP-1 are released in response to the active absorption of glucose, which must be present in the gut at levels above fasting values, before either hormone can be released.[9] It is likely, therefore that the attenuation seen in GIP and GLP-1 release is due to reductions in the rate of glucose absorption.

The reputed health benefits of a diet rich in complex carbohydrates are numerous. The soluble fibres, such as oat gum, guar gum and pectin are thought to be important in reducing plasma low density lipoprotein cholesterol levels.[10] It has been suggested that an increased intake of such fibres may help reduce the risk of coronary heart disease. There is also an emerging link between hyperinsulinaemia and cardiovascular disease[11] and therefore an increased intake of fibres which delay glucose absorption and the subsequent release of gut hormones such as insulin, GIP and GLP-1 may be beneficial.

As can be seen from this study, it is possible to incorporate concentrated forms of guar galactomannan into palatable food products, which have positive effects on glycaemic control. Development of such foodstuffs, may provide a way of including concentrated forms of s-NSP, into the habitual diet.

5.1 Future Work

Blood samples have been taken for insulin analysis, at the same time points as those for glucose, GIP and GLP-1. The trial is still ongoing and once sufficient recruits have passed through the trial, glucose and gut hormone data is to be correlated with the viscosity data. A better understanding of the structure-function relationship of guar may allow a more precise development of a clinically effective functional food product.

6 REFERENCES

1. P.R. Ellis, T.Kamalanathan, F.M. Dawoud, R.N. Strange and T.P. Coultate, *Euro. J. Clin.Nut.*, 1988, **42**, 425.
2. L.M. Morgan, J.A.Tredger, J.Wright and V.Marks, *Brit. J. Nut.*, 1990, **64**, 103.
3. D.J.A.Jenkins, A.R.Leeds, M.A.Gassull, B.Cochet and K.G.M.M. Alberti, *Annals on Internal Medicine*, 1977, **86**, 20.
4. V.E.Morris, "Chemical and Biological Aspects of Dietary Fibre", Ed. D.A.T. Southgate, Royal Society of Chemistry, London, 1990, p. 91.
5. F.G.Roberts, "Chemical and Biological Aspects of Dietary Fibre", Ed. D.A.T. Southgate,
   Royal Society of Chemistry, London, 1990, p. 164.
6. R.M.Fairchild, P.R.Ellis, A.J.Byrne, S.D.Luzio and M.A.Mir, *Brit. J. Nut.*, 1996, **76**, 63.
7. H.S. Fuessl, *Practical Diabetes*, 1986, **3**, 258.
8. P.R. Ellis and E. R. Morris, *Diabetic. Med.*, 1991, **8**, 378.
9. L.M. Morgan, "Nutrient regulation of insulin secretion", Ed. P.R. Flatt, Portland Press, London, 1992, p. 1.
10. J.W. Anderson, *Can. J. Cardiol.*, 1995, **11**, 55G.
11. F. Defronzo and E. Ferrannini, *Diabetes care*, 1991, **14**, 173.

OATS—A MULTIFUNCTIONAL FOOD

R.W. Welch

Northern Ireland Research Centre for Diet and Health
University of Ulster
Coleraine BT52 1SA

1 INTRODUCTION

This review outlines current consumption levels and the nutritive value of oats are also described. The physiological effects which qualify oats as a multifunctional food are also described. Effects on cholesterol metabolism, glucose metabolism and on gut function are well established. In addition there are a number of potential effects of dietary oats which include promoting weight loss, alleviating hypertension, enhancing physical performance and mood, improving immune status and enhancing antioxidant protection.

2 OAT CONSUMPTION

Oats was formerly a dietary staple in some Northern and Western regions of Europe.[1] However current oat consumption levels are low in all parts of the world, reaching a maximum of 6-8g/day in Scandinavia.[2] At this level oats comprises only 3-4% of total cereal intake and ~1% of total energy intake.

3 NUTRITIVE VALUE

Oats is a whole kernel cereal most commonly consumed as groats (kernels) milled to yield various products (rolled oats, oatmeal, oat flour). Although there are variations depending on the source of the raw material and on subsequent processing, the nutritional profile of oat groats compares favourably with other unfortified cereals.[1] Oats has a higher oil content and thus a higher energy density, oat protein has a relatively good amino acid balance, and oats also contains significant amounts of dietary fibre, which is high in soluble fibre.[3] The soluble fibre, which is comprised mainly of β-glucan, is also known as oat gum.

Oats also contributes significant amounts of dietary minerals (Mg, P, Fe, Cu, Zn) and vitamins (thiamin, vitamin E, folate, niacin).[1] In addition to fibre and nutrients, oats also contains a wide range of non-nutrient compounds of potential physiological significance.[1]

Other oat products are available. These include oat bran which is a coarse milling fraction, with a higher total fibre, soluble fibre and protein content than the original groats. Since the soluble fibre appears chiefly responsible for many of the physiological effects of oats, this product has received increasing attention in recent years.

4 OATS AS A MULTIFUNCTIONAL FOOD

Dietary oats has been shown to confer a number of significant physiological effects in the prevention or alleviation of disease, and thus may be considered as a multifunctional food. Established effects comprise improvements in gut function, reductions in plasma cholesterol, and modulation of plasma glucose levels. Oats may also impact favourably on other physiological processes of significance in the prevention or alleviation of disease. Established and potential effects of dietary oats are outlined below, and summarised in Table 1.

4.1 Established Effects

4.1.1 Improving Gut Function. Fibre rich foods are known to exert a number of effects on gut function. These include reductions in transit time and increases in faecal bulk.[4] While these effects are valuable in the short term as an aid to laxation, in the longer term they may also be beneficial in the prevention of other diseases such as colon cancer. Earlier work has shown that inclusion of oats in the diet can increase fecal bulk and defecation frequency, and reduce transit time.[1] A recent study has shown that oat fibre exerts similar effects to wheat fibre.[5] Fecal bulk was increased and transit reduced, and there were also reductions in faecal pH which may be beneficial to gut health.

4.1.2 Reducing Plasma Cholesterol. The ability of oats to reduce plasma cholesterol, and in particular the low density lipoprotein (LDL)-cholesterol fraction, has received substantial attention. The soluble fibre β-glucan gum is the major hypocholesterolaemic component.[1,6,7] Evidence for this effect has not been unequivocal,[1,8] and the underlying mechanisms are not fully understood.[1,9] However the United States Food and Drug Administration (FDA) has recently recognised a claim which states *'diets low in saturated fat and cholesterol that include soluble fibre from whole oats may reduce the risk of heart disease'.*[8] In permitting this claim, the FDA accepted that the causes of heart disease are multifactorial, but that there is good evidence that raised plasma cholesterol is a major risk factor. To be eligible for a claim, the FDA ruling requires oat products to provide at least 3g β-glucan/day which is equated to 60g whole oat product (flakes, meal or flour) or 40g oat bran.[8]

4.1.3 Reducing Glycaemia. Earlier studies on the effects of oats on cholesterol metabolism indicated that diabetic control was improved.[1] In comparison with many other foodstuffs, the postprandial glycaemic response of oats is low.[10] A number of factors contribute to this effect but a major factor is the viscous β-glucan gum.[11] Recent, longer term studies with oat products also show reduced glycaemia which is associated with reduced insulinaemia.[12,13] Animal studies indicate that oats can impact on gut hormones,[14] which may be relevant to the effects on both glucose and cholesterol metabolism, and of significance to gut health.

4.2. Potential Effects

4.2.1 Inducing Satiety and Promoting Weight Loss. Satiety, which is under complex neurological and humoral control, is important in the regulation of food intake. Foods which induce increased satiety may be useful for limiting food intake and thus alleviating or preventing obesity and associated disorders such as diabetes. A recent study has shown that a novel liquid oat product, which increased satiety, led to substantial weight loss when used in a weight control study. There were also concomitant reductions in blood glucose and insulin levels.[15]

4.2.2 Alleviating Hypertension. Hypertension is associated with hyperinsulinaemia, and thus may be expected to respond to reductions in plasma insulin levels. Although earlier studies gave equivocal results, a recent trial with hypertensives showed that dietary oats led to reductions in both systolic and diastolic blood pressure which were associated with improvements in plasma lipid profiles and post-prandial insulin levels.[16]

4.2.3 Preventing Coronary Heart Disease. Elevated plasma cholesterol is only one risk factor in heart disease. There are no reports that directly link oat consumption to either atherosclerosis or coronary heart disease in humans. However hyperinsulinaemia has been implicated in the proliferation and migration of arterial smooth muscle cells.[17] Thus the reduced insulinaemia resulting from oat consumption may augment the effects of lowered cholesterol in the prevention or alleviation of atherosclerosis. Furthermore, in an animal study, oats was found to reduce the incidence of aortic rupture in turkeys fed the growth hormone diethylstilboestrol.[18]

4.2.4 Safety for Coeliac Patients. Oats has been proscribed in the diets of patients with coeliac disease, since the discovery of this disease. However oats is not closely related to the other cereals (wheat, rye, barley) implicated in coeliac disease, and two recent studies have shown that consumption of significant amounts of oats did not lead to any adverse effects in either newly diagnosed coeliac patients,[19] or coeliac patients in remission.[19,20] The longest of these studies ran for 12 months. However the presence of coeliac-inducing gliadins in oats requires further study,[21] and additional longer term trials are needed to ensure the safety of oats for coeliac patients, who also need to be reassured that oats can be supplied that is free of contamination with other cereals.

4.2.5 Improving Gut Health. In addition to the established effects of oats on gut function described above, there are a number of other potential ways in which oats may improve gut health. These include prebiotic effects of oat products. Prebiotics are fermentable substrates that influence the spectrum of the gut microflora and the gut fermentation characteristics in a number of beneficial ways.[22] Although there is as yet no direct data available for oats, oats provide a range of substrates, including soluble fibre, for large bowel fermentation.[23] Other prebiotic food components increase beneficial species, and the effects include inhibiting the pathogen growth, lowering gas production, metabolising xenobiotics, reducing tumour formation, and stimulating gut immunological processes.[22] Furthermore fermented oat foodstuffs have been developed which exert beneficial effects on the intestinal mucosa.[24] Such effects may be due to prebiotic effects as outlined above, to a probiotic effect of the fermenting organisms, or to effects on the enterocytes, mediated by β-glucan, and which are associated with changes in gut hormones.[14]

Table 1. *Summary of Established and Potential Physiological Effects of Dietary Oats*

EFFECTS	MECHANISMS/AGENTS
Established	
Improving gut function	increased fecal bulk; reduced transit
Reducing plasma total and LDL-cholesterol	soluble fibre (β-glucan); increased bile acid excretion etc
Reducing glycaemia and insulinaemia	reduced gastric emptying; reduced absorption; viscosity effect
Potential	
Inducing satiety and weight loss	*? reduced gastric emptying*
Alleviating hypertension	*reduced glycaemia and insulinaemia*
Preventing coronary heart disease	*lowered cholesterol; ? reduced atherosclerosis; ? antioxidant protection*
Safety for coeliac patients	*? absence of toxic gliadin peptides*
Improving gut health	*laxation; proliferative effects; ? prebiotic effects; ? probiotic effects*
Enhancing performance and mood	*? reduced glycaemia and insulinaemia; ? psychotropic effects*
Cancer protectant	*gut effects - bulking, transit; ? prebiotic effects; ? phytoestrogens; ? antioxidant protection*
Enhancing immune status	*? prebiotic effects in gut; ? immunostimulants*
Enhancing antioxidant status	*nutrient antioxidants; ? non-nutrient antioxidants*

4.2.6 Enhancing Physical Performance and Mood. Oats is traditionally associated with an image of vigour and wellbeing. A number of early studies indicated that oats may improve athletic and endurance performance, and may impact favourably on psychological factors.[25] Recent evidence has shown that oats, in comparison with other cereals, improved glucose and insulin status during exercise but there were no effects on performance.[26] Morphine antagonists have been reported in oats,[27] and peptides with opioid activity have been found in tryptic digests of wheat.[28] Thus it is possible that oats possesses unidentified components with the potential to influence psychological factors.

4.2.7 Enhancing Immune Status. The potential stimulation of gut immunological processes via a prebiotic effect of oats is outlined above.[22] There are other potential ways in which oats could impact on immune status. β-glucans from fungal sources can potentiate various aspects of the immune system.[29] Although oat β-glucans do not appear to possess the necessary structure for immunoactivity,[29] barley β-glucan, which is structurally similar to that in oats, is reported to affect immune responses.[30] In addition, avenanthramides, a class of phenolic antioxidants found in oats, are structurally similar to a number of compounds which possess pharmaceutical antiallergic activity.[31]

4.2.8 Cancer Protectant. The effects of oats on gut function and health outlined above may help reduce the onset of colon cancer by a number of mechanisms. Additional cancer protection could be afforded by phytoestrogens, which are putative natural cancer protectants that include diflavonoids and lignans.[32] The sterol, β-sitosterol, occurs in oats and is reported to have phytoestrogen activity.[33] There are no confirmed reports of other phytoestrogens in oats, but they have been identified in other cereals.[32] Antioxidants may also be important in cancer onset and progression,[34] and the potential role of oat antioxidants in such processes is outlined below.

4.2.9 Enhancing Antioxidant Status. Oxidant damage to biological molecules brought about by reactive oxygen species may be an important factor in the onset and progression of a number of disease states including heart disease and cancer.[34,35] There is an array of endogenous enzymic and non-enzymic antioxidant defence mechanisms which limit or restore this damage.[34,35] Nutrients are important in a number of these mechanisms (eg Vitamin E, Cu) and non-nutrient antioxidants have also been implicated in beneficial effects. For example, the intake of non-nutrient flavonoids has been correlated with reduced mortality from coronary heart disease.[36] Oats contains a wide range of antioxidants, and oat flour has been used as a food antioxidant.[2] Oat antioxidants include tocols, sterols, phenolics, phytate and saponins. The physiological significance of oat antioxidants has not been investigated. However they have the potential to alleviate damage by reactive oxygen species and thus may play a role in the prevention or alleviation of coronary heart disease, cancer and other diseases.

5 SUMMARY AND CONCLUSIONS

Although oats has a comparatively good nutritional profile, current consumption levels are low. Oats have been shown to confer a number of beneficial effects in the diet. Established effects include improving gut function, reducing plasma cholesterol levels and in particular LDL cholesterol, and modulating plasma glucose levels. In addition oats may confer other beneficial effects including alleviating hypertension, enhancing physical performance and mood, improving immune status and increasing antioxidant protection. Further work is required to fully elucidate the mechanisms underlying the established benefits, and to evaluate the potential effects of this cereal.

6 REFERENCES

1. R.W. Welch, *Oats in human nutrition and health*, in The Oat Crop - Production and Utilisation (ed. R.W. Welch), Chapman & Hall, 1995, London, 433-79.
2. Food Consumption Statistics 1979-1988, Organisation for Economic Co-operation and Development, 1991, Paris.
3. R.W. Welch, *The chemical composition of oats*, in The Oat Crop - Production and Utilisation (ed. R.W. Welch), Chapman & Hall, 1995, London, 279-320.
4. C.Hillemeier, Pediatrics, 1995, 96, 997-9.
5. K.B.Hosig, F.L.Shinnick, M.D.Johnson, J.A.Story & J.A.Marlett, Cereal Chem.,

1996, **73**, 392-8.

6. C.M.Ripsin, J.M.Keenan, D.R. Jacobs, P.J. Elmer, R.W. Welch, and 11 others, J. Amer. Med. Assoc., 1992, **267**, 3317-25.

7. J.T.Braaten, F.W.Scott, P.J.Wood, K.D.Reidel, M.S.Wolynetz, D.Brule & M.W.Collins, Diabet. Med., 1994, **11**, 312-18.

8. United States, Food and Drug Adminstration, Federal Register, 1996, **61**, 296-339, and L-S Document 582663, Jan 23 1997.

9. J.A.Marlett, K.B.Hosig, N.W.Vollendorf, F.L.Shinnick, V.S.Haack & J.A.Story, Hepatol., 1994, **20**, 1450-7.

10. T.M.S.Wolever, L.Katzman-Relle, A.L.Jenkins, V.Vuksan, R.G.Josse & D.J.A.Jenkins, Nutr. Res., 1994, **14**, 651-9.

11. P.J.Wood, J.T.Braaten, F.W.Scott, K.D.Reidel, M.S.Wolynetz & M.W.Collins, Br. J. Nutr., 1994, **72**, 731-43.

12. L.Tappy, E.Gügolz, P.Würsch, Diab. Care., 1996, **19**, 831-4.

13. M.E.Pick, Z.J.Hawrysh, M.I.Gee, E.Toth, M.L.Garg & R.T.Hardin, J.Am. Diet. Assoc., 1996, **96**, 1254-61.

14. E.K.Lund, S.R.R. Musk, & I.T. Johnson, (1993) Proc. Nutr. Soc. 1993, **52**, 124A

15. E.Rytter, C.Erlanson-Albertsson, L.Lindahl, I.Lundquist, U.Viberg, B.Åkesson & R.Öste, Ann. Nutr. Metab. 1996, **40**, 212-20.

16. J.Keenan, 1997, personal communication.

17. R.W.Stout, Lancet, 1987, i, 1077-9.

18. C.F.Simpson & R.H.Harms, Poult. Sci. 1969, **48**, 1757-61.

19. E.K.Janatuinen, P.H.Pikkarainen, T.A.Kemppainen, V-M. Kosma, R.M.K.Järvinen, M.I.J.Uusitupa & R.J.K.Julkunen, N. Engl. J. Med., 1995, **333**, 1033-7.

20. U. Srinivasan, N.Leonard, E.Jones, D.D.Kasarda, D.G.Weir, C.O'Farrelly & C.Feighery, Brit. Med. J., 1996, **313**, 1300-1.

21. R.Troncone, L.Maiuri, A.Leone, G.Mazzarella, F.Maurano, L.Vacca, C.Cucci, M. De Vincenzi & S.Auricchio, J. Pediatr. Gastroenterol. Nutr., 1996, **22**, 414.

22. G.R.Gibson, A.Willems, S. Reading & M.D.Collins, Proc. Nutr. Soc., 1996, **55**, 899-912.

23. Å.Lia, B.Sundberg, P. Åman, A-S, Sandberg, G. Hallmans & H. Andersson, Brit. J. Nutr., 1996, **76**, 797-808.

24. M.L.Johannsson, G.Molin, B.Jeppsson, S.Nobaek, S.Ahrne & S.Bengmark, App. Env. Microb., 1993, **59**, 15-20.

25. J.Kühnau, J. & W. Ganssmann, *Oats an Element of Modern Nutrition*, Umschau Verlag, 1985, Frankfurt am Main.

26. G.L.Paul, J.T.Rokusek, G.L.Dykstra, R.A.Boileau & D.K.Layman, J. Nutr.,1996, **126**, 1372-81.

27. J.Connor, T.Connor, P.B.Marshall, A.Reid & M.J.Turnbull, J. Pharm. Pharmacol., 1975, 27, 92-8.

28. C. Zioudrou, R.A.Streaty, & W. A.Klee, J. Biol. Chem. 1979, **254**, 2446- 49.

29. J.A.Bohn & J.N.BeMiller, Carb.Polymers. 1995, **28**, 3-14.

30. A.Estrada, B.Li., M.Redmond & B.Laarveld, Faseb J., 1994, **8**, A757.

31. F.W.Collins, J. Agric. Fd. Chem., 1989, **37**, 60-6.

32. H.Adlercreutz, Env. Hlth. Persp., 1995, **103**, 103-12.

33. E.R.Rosenblum, R.E.Stauber, D.H.Van Thiel, I.M.Campbell & J.S.Gavaler, Alcohol Clin. Exp.Res.,1993, **17**, 1207-9.

34. M.L.Burr, J.Hum. Nutr. Dietet., 1994, **7**, 409-16.
35. R.M.Hoffman & H.S.Garewal, Arch. Int. Med., 1995, **155**, 241-6.
36. M.G.L.Hertog, E.J.M.Feskens, P.C.H.Hollman, M.B.Katan & D.Kromhout, Lancet, 1993, **342**, 1007-11.

SHORT-TERM CONSUMPTION OF BRUSSELS SPROUTS LEADS TO AN INCREASE IN DETOXIFYING GLUTATHIONE-S-TRANSFERASE ENZYMES AND TO A DECREASE IN OXIDATIVE DNA-DAMAGE IN HUMAN VOLUNTEERS

H. Verhagen and G. van Poppel

TNO Nutrition and Food Research Institute
PO Box 360
3700 AJ Zeist
The Netherlands

1 INTRODUCTION

Epidemiological studies indicate that the consumption of fruits and vegetables is negatively associated with the development of degenerative diseases such as cancer and coronary heart diseases. The putative beneficial effect of brassica species in particular is supported by the results of many case-control studies (Tables 1 and 2), unlike cohort studies in which this effect has not been established. Two important mechanisms underlying anticarcinogenic potential are antioxidant potential and alteration of biotransformation capacity. In humans, the former can be assessed by measuring the levels of 8-oxo-7,8-dihydro-2'-deoxyguanosine (8-oxodG) in urine. The latter can be assessed for example by the induction of the phase-II biotransformation enzymes glutathione-S-transferases (GSTs). In two short-term intervention studies with Brussels sprouts we have investigated the hypothesis that the putative beneficial effects of brassica vegetables in humans are due to induction of detoxification enzymes as well as to antioxidant properties.

2 METHODS

The study designs are shown in Table 3. In the control period, the volunteers refrained from any cruciferous vegetables and consumed 300g cooked non-brassica vegetables/day. During the intervention periods volunteers consumed 300g cooked Brussels sprouts/day. On days 13 and 14 (first study) and day 7 (second study) of both periods blood samples were collected. Urine was collected on day 12 (first study) and day 7 (second study) of both periods over a 24-hour period. In addition, in the second study duodenal and rectal biopsies were taken on day 7 of both periods.

3 RESULTS

In the first intervention study in the control group, similar GST-α levels were found in both periods ($P = 0.814$), while in the Brussels sprouts group the GST-α levels were elevated by a factor 1.4 ($P = 0.002$) (Figure 1). In the control group there was

Table 1 *Summary of the results of case-control studies investigating an association between consumption of brassica vegetables and cancer development; data shown for different vegetables* (Verhoeven *et al.* 1996)[5]

Type of brassica	Number of studies showing an association (of which statistically significant)			Number of studies showing a different association per gender (inv: inverse; pos: positive; no:no association)	Total number of studies
	Inverse association	No association	Positive association		
All brassicas	39 (20)	4	9 (1)	1:♀ inv./♂ pos. 1:♀ no/♂ pos. 4:♀ pos./♂ inv.	58
Cabbage (excluding chinese cabbage)	17 (7)	5	2 (1)		24
Broccoli	10 (7)	3	1 (0)	1:♀ inv./♂ no 2:♀ no/♂ inv. 1:♀ pos./♂ no	18
Cauliflower	8 (2)	1	2 (0)	1:♀ pos./♂ no	12
Brussels sprouts	2 (0)	2	2 (0)	1:♀ pos./♂ inv.	7

Table 2 *Summary of the results of case-control studies investigating an association between consumption of brassica vegetables and cancer development; data shown for different cancer types* (Verhoeven *et al.* 1996)[5]

Cancer site	Number of studies showing an association for one or more brassica vegetables (of which statistically significant)		Total number of studies
	Inverse association	Positive association	
Colon	11 (6)	4 (0)	15
Stomach	8 (5)	3 (1)	11
Rectum	8 (4)	3 (0)	10
Lung	9 (6)	2 (0)	9

Table 3 *Study Designs for the First and Second Intervention Study with Brussels Sprouts*

Study 1: 10 males; 300 g vegetables per day		
3 weeks	control diet (n = 5)	control diet (n = 5)
3 weeks	control diet (n = 5)	Brussels sprouts (n = 5)
Study 2: 5 males, 5 females; 300 g vegetables per day		
1 week	Brussels sprouts (n = 5)	control diet (n = 5)
1 week	control diet (n = 5)	Brussels sprouts (n = 5)

Table 4 *Second study: levels of 8-oxodG in urine (pmol/kg/24hours) on day seven of the control period and on day seven of the Brussels sprouts period*

Males			Females		
Volunteer number	Control period	Brussels sprouts period	Volunteer number	Control period	Brussels sprouts period
2	448	416	1	433	165
4	1358	5427	3	393	804
6	772	525	5	165	334
8	1469	533	7	688	351
10	405	236	9	113	446

no difference between the two periods in levels of 8-oxodG. In contrast, in the Brussels sprouts group the levels of 8-oxodG were decreased by 28% during the intervention period (P=0.039) (Figure 2).

In the second intervention study near the end of the Brussels sprouts period, a significant increase (1.5-fold) in plasma GST-α levels was observed in males but not in females ($P = 0.031$) (Figure 3). As a result of the dietary regimen rectal GST-α and GST-π levels were slightly increased at the end of the Brussels sprouts period, by 30 and 15% respectively, (P<0.05 and P<0.005) (Figure 4). In four of five males a reduction in 8-oxodG was found, whereas in one male (volunteer no. 4) the 8-oxodG excretion was unusually high in the control period and was considerably higher in the Brussels sprouts period. In females no effect of consumption of Brussels sprouts on excretion of 8-oxodG was found (Table 4).

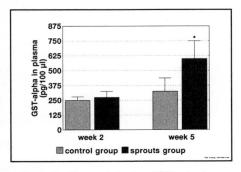

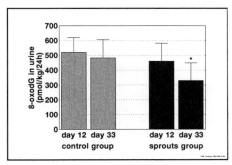

Figure 1 *First study levels: of GST-α in plasma (pg/100μl) on days 13/14 of the control and Brussels sprouts period*

Figure 2 *First study: levels of 8-oxodG urine (pmol/kg/24h) on day 12 of the control and of the Brussels sprouts period*

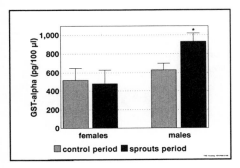

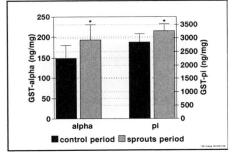

Figure 3 *Second study: levels of GST-α in plasma (pg/100μl) on day seven of the control and Brussels sprouts period*

Figure 4 *Second study levels of GST-α and GST-π in rectal biopsies (ng/mg) on day seven of the control and Brussel sprouts period*

4 DISCUSSION

The present data lend further support to the possibility that brassica vegetables may have health promoting effects under physiologically feasible conditions and without apparent side effects, as apparent from an induction of the detoxification enzymes GST-α and GST-π and by a decrease in the rate of oxidative DNA-damage. Our experimental finding that males rather than females seem to be susceptible to the effects of Brussels sprouts deserves further study. Since brassica vegetables differ from other vegetables by the presence of glucosinolates these health promoting effects seem to be attributable to one or more of these phytochemicals or to their breakdown products. In addition, these studies demonstrate the possibilities of establishing the efficacy of functional foods and food ingredients efficiently and directly via short-term intervention studies with human volunteers.

5 ACKNOWLEDGEMENTS

We appreciate the contributions made to this work by our colleagues whose names can be found in the references provided below. These references provide full details of the two intervention studies.

6 REFERENCES

1. J.J.P. Bogaards *et al.*, *Carcinogenesis* 1994, **15**, 1073.
2. H.Vergahen *et al.*, *Carcinogenesis* 1995, **16**, 969.
3. W.A. Nijhoff *et al.*, *Carcinogenesis* 1995, **16**, 955.
4. W.A. Nijhoff *et al.*, *Carcinogenesis* 1995, **16**, 2125.
5. D.T.H. Verhoeven *et al.*, *Cancer Epid., Biom. Prev.* 1996, **5**, 773.
6. M.N. Nanjee *et al.*, *Am.J.Clin.Nutr.* 1996, **64**, 706.
7. H.Verhagen *et al.*, *Cancer Letters* 1997, in press.
8. D.T.H. Verhoeven *et al.*, *Chem-Biol. Interact.* 1997, **103**, 79.

CURCUMIN ETHANOL-AQUEOUS EXTRACT INHIBITS *IN VITRO* HUMAN LOW DENSITY LIPOPROTEIN LIPOPEROXIDATION

M.C. Ramírez-Tortosa,[1] C.M. Aguilera,[1] M.A. Carrión-Gutiérrez,[2] A .Ramirez-Boscá[2] and A.Gil[1]

[1]Department of Biochemistry and Molecular Biology, Institute of Nutrition and Food Technology, University of Granada, Granada, Spain
[2]ASAC Medical Department, Alicante, Spain

1 INTRODUCTION

Curcumin (diferuloylmethane), a yellow pigment obtained from rhizomes of *Curcuma longa,* is a major component of turmeric which is commonly used as a spice and food-colouring agent. Other minor compounds, namely demethoxy- and bis-demethoxycurcumin and the protein turmerin are also present in turmeric extracts. Curcuma extracts have been shown to exhibit a number of functions. It has been reported that administration of pure or commercial grade curcumin in the diet decreases the incidence of tumours in mice and also reduces tumour size. Histopathological examination of the tumours showed that dietary curcumin inhibits the number of papillomas and squamous cell carcinomas of the forestomach as well as the number of adenomas and adenocarcinomas of the duodenum and colon.[1,2]

The effect of curcumin on different stages of skin cancer development has been studied. The results show that curcumin inhibits cancer development at initiation, promotion and progression stages.[3] In addition, curcumin produces an anti-inflammatory action and inhibits vascular smooth muscle cell proliferation.[4]

Previous reports have shown that the curcuma extract used in this study has a powerful antioxidant effect on the peroxidation of unsaturated fatty acids and animal organ homogenates *in vitro* as well as decreasing blood lipid peroxide levels in humans[5] and mice.[6] Intake of the hydroalcoholic turmeric extract by healthy humans for a period of 45 days resulted in a significant decrease of high density lipoprotein (HDL) and low density lipoprotein (LDL) peroxidation.[5] In mice, administration of the extract inhibited *in vitro* lipid peroxidation of liver homogenates.[6]

Oxidation of LDL plays a pivotal role in the pathogenesis of atherosclerosis therefore treatment with antioxidants may prevent the development of coronary heart or peripheral arterial diseases.[7-9] The aim of this study was to evaluate the antioxidant effect of one ethanol-aqueous extract obtained from the rhyzome of *Curcuma longa* on the Cu^{2+}-mediated oxidative modification of human LDL.

2 MATERIALS AND METHODS

The hydroalcoholic extract from the rhyzome of *Curcuma longa* (with a 10% concentration of curcumin) was provided by A.S.A.C. Pharmaceutical International A.I.E. (Alicante).

Human blood (25ml) was collected by venipuncture into EDTA-containing vacutainer tubes after a 12 hour overnight fast. Samples were kept on ice and centrifuged at 1700 x g for 15 minutes at 4 °C to obtain plasma.

2.1 Lipoprotein Isolation.

Very low density lipoprotein (VLDL), LDL and HDL were isolated by a single discontinuous density gradient ultracentrifugation in a vertical rotor.[10] A discontinuous NaCl/KBr density gradient was formed by adjusting the density of the plasma to 1.30g/ml with KBr and layering normal saline solution (d=1.006g/ml) over the adjusted plasma. The isolated LDL was exhaustively dialyzed against 150mN NaCl, 1mM EDTA pH 7.4 at 4°C for 24 hours.

2.2 Determination of LDL Oxidation Susceptibility

2.2.1 Thiobarbituric acid reactive substances determination. LDL (200µg/ml) was oxidized in the presence of copper (Cu^{2+}) (1.25µM, 2.5µM, 5µM, 10µM and 20µM) in phosphate buffered saline (PBS) for 24 hours at 37°C. After incubation the samples were placed on ice and the oxidation terminated by the addition of 200µM EDTA and 40µM butylated hydroxytoluene.[11]

The lipid peroxide content of the oxidized LDL was determined as thiobarbituric acid reactive substances (TBARS) as described by Buege and Aust.[12] Oxidized LDL was vortexed with 1ml trichloroacetic acid-thiobarbituric acid-hydrochloric acid solution (15% w/v TCA; 0.37% TBA, 25 mol/L HCl) and heated for 20 minutes at 100 °C to precipitate protein. After cooling, samples were centrifuged at 1700 x g for 20 minutes. The supernatant was carefully removed and the absorbance determined at 532nm against a blank containing all the reagents but minus LDL. The absorbance was expressed in malondialdehyde (MDA) equivalents/mg LDL protein using a standard curve for 1,1,3,3,tetramethoxypropane. Absorbance versus Cu^{2+} concentration curves were plotted and the lag phase and slopes calculated using Slidewrite software.

2.2.2 Conjugated dienes. Conjugated diene formation was determined as described by Esterbauer et al.[13] LDL protein (20µg/ml) was incubated with increasing concentrations of hydroalcoholic turmeric extract (2.4µg, 4.8µg and 9.6µg curcumin) and 5µM Cu^{2+} in a 1ml quartz cuvette at 37 °C. Absorbance was read at 234nm every 10 minutes for a maximun of three hours using a Perkin Elmer Spectrophotometer . The results are shown in nmol conjugated diene/mg LDL protein.

3 RESULTS

To compare the antioxidant effects of the turmeric extract with that of vitamin E on

LDL oxidation, LDL was incubated with $CuSO_4$ and increasing concentrations of extract or dl-α-tocopherol. As shown in Table 1, the presence of turmeric led to an inhibition of LDL oxidation when Cu^{2+} concentrations were low (1.25μM). Neither curcuma extract nor tocopherol were able to inhibit LDL oxidation at Cu^{2+} concentrations higher than 2.5μM after 24 hour incubation.

The turmeric extract decreased conjugated diene formation parallel to the curcumin concentration in comparison with the control group. Values of conjugated diene were higher than those obtained using tocopherol as the antioxidant agent (Table 2).

The effect of increasing concentrations of tumeric extract on the lag phase of conjugated diene formation in LDL incubated *in vitro* with 5μM Cu^{2+} for 160 minutes is shown in Table 3. The increase in lag oxidation phase with increasing turmeric extract concentration indicated that the extract inhibited LDL oxidation at the initial stage.

4 DISCUSSION

The results presented in this study show that the ethanol-aqueous extract of turmeric increases human LDL resistance to *in vitro* oxidation. It was found that when LDL was submitted to $CuSO_4$ mediated oxidation, the addition of extract to the incubation medium increased the lag phase of conjugated diene formation and decreased the formation of total conjugated dienes because of an inhibition in the initial oxidation process.

Our results show that the extract of turmeric is unable to stop the LDL oxidation process once it has been initiated. This effect is in agreement with the inability of flavonoids to stop LDL oxidation in more advanced stages of the oxidation process.[14] It has been shown that curcumin treatment of human subjects decreases serum total cholesterol as well as (LDL+VLDL) cholesterol, serum lipoprotein (LDL+VLDL) concentrations and increases HDL cholesterol.[15] In addition, it decreases the generation

Table 1 *TBARS concentration in human LDL after 24 hours incubation with increasing concentrations of Cu^{2+} in the presence of turmeric extract or dl-α-tocopherol*

Copper concentration	nmoles TBARS/mg LDL Protein				
	Control	Turmeric extract 3.3μg	Turmeric extract 1.5μg	dl-α-tocopherol 3.3μg	dl-α-tocopherol 1.5μg
1.25μM	6.06	3.64	4.99	7.79	3.01
2.5μM	6.15	7.6	8.27	6.34	10.39
5μM	53.27	51.8	57.13	54.72	48.76
10μM	63.87	70.52	76.5	70.81	56.74
20μM	73.03	76.4	74.1	75.72	71.1

Table 2 *Formation of conjugated dienes in LDL incubated in vitro with 5μM Cu^{2+} and increasing concentrations of turmeric extract or dl-α-tocopherol*

		pmoles conjugated diene/mg LDL protein					
		Turmeric extract (μg/ml)			Tocopherol (μg/ml)		
Minutes	**Control**	**2.4**	**4.8**	**9.6**	**2.4**	**4.8**	**9.6**
10	0	0	0	0	0	0	0
20	0	0	0	0	0	0	0
30	6	3	0	0	1	0	0
40	8	5	5	2	1	0	0
50	12	8	6	3	2	3	2
60	14	8	8	5	2	3	3
70	22	12	9	6	8	7	3
80	30	17	9	8	9	8	3
90	58	34	25	19	24	9	3
100	70	29	25	20	26	9	6
110	100	47	33	18	47	9	6
120	111	62	50	27	61	12	7
130	123	78	56	29	92	24	7
140	125	101	78	44	101	28	13
150	129	119	103	48	107	38	19
160	132	120	111	60	111	52	25

Table 3 *Effect of increasing concentrations of turmeric extract or dl-α-tocopherol on the lag phase of conjugated diene formation in LDL incubated in vitro with 5μM Cu^{2+} for 160 minutes*

	Lag phase (minutes)		
	Control	**Turmeric Extract**	**Tocopherol**
0μg	62.77	62.77	62.77
2.4μg	-	76.75	84.09
4.8μg	-	94.71	119.74
9.6μg	-	137.18	295.78

of reactive oxygen species in rat peritoneal macrophages[16] and inhibits platelet aggregation probably because of an alteration in the eicosanoid metabolism in human blood platelets[17] and decreases the blood lipid peroxide levels in humans. These effects of turmeric extracts on plasma lipids may contribute to decelerate the progression of cardiovascular diseases. Atherosclerosis, a complex disease of the artery wall, is associated with the accumulation of macrophages derived from cells laden with a modified form of LDL not recognised by native LDL receptors.[18, 19] Possible biological mechanisms leading to LDL modification include aggregation and oxidation, either of which could lead to uptake by macrophages through scavenger receptors, phagocytosis or receptors specific to oxidised LDL.[20, 21] The extract of turmeric may interfere with the development of atherosclerosis because of its influence in limiting the susceptibility of LDL oxidation.

5 REFERENCES

1. M.A. Azuine, S.V. Bhide, *J. Ethnopharmacol*, 1994, **44**, 211.
2. M.T. Huang, Y.R. Lou, W. Ma, H.L. Newmark, K.R. Reuhl, A.H. Conney, *Cancer Res.*, 1994, **54**, 5841.
3. M. Nagabhushan, S.V. Bhide, *J. Am. Coll. Nutr.* , 1992, **11**, 192.
4. M. Susan, M.N. Rao, *Arzneimittelforschung*, 1992, **42**, 962.
5. A. Ramirez-Bosca, A. Soler, M.A. Carrion, J. Laborda, E. Quintanilla, *Age*, 1995, **18**, 167.
6. J. Miquel, M. Martinez, A. Diez, E. DeJuan, A. Soler, A. Ramirez, J. Laborda, M. Carrion, *Age*, 1995, **18**, 171.
7. G.E. Fraser, *Am J. Clin. Nutr.* 1994, **59**, 11175.
8. H. Esterbauer, M. Rotheneder, G. Striegl, A. Ashy, W. Satytler, G. Jurgen., *Fat. Sci.Technol.*, 1989, **91**,316.
9. M. Suzukawa, M. Abbey, P.R. Howe, P.J. Nestel. *J. Lipid Res.* 1995, **36**, 473.
10. B.H. Chung, T. Wilkinson, J.C. Geer, J.P. Segrest. *J. Lipid Res.*, 1981, **21**, 284.
11. I. Jialal, S.M. Grundy., *J. Clin Invest*, 1991, **87**, 597.
12. J.A. Buege, S.D. Aust. *Methods Enzimol.*, 1978, **52**, 302.
13. H. Esterbauer, G. Striegl, H. Puhl, S. Oberreither, M. Rotheneder, M. El-Saadani, G. Jurgens *Ann N. Y. Acad. Sci.*, 1989, **570**, 254.
14. M. Viana, C. Barbas, B. Bonet, M.V. Bonet, M. Castro, M.V. Fraile, E. Herrera. *Atherosclerosis*, 1996, **123**, 83.
15. G.S. Seetharamaiah, N. Chandrasekhara, *J. Food Sc. Techn. Mysore*, 1993, **30**, 249.
16. B. Joe, B.R. Lokesh, *Biochim Biophys Acta*, 1994, **1224**, 255.
17. K.C. Srivastava, A. Bordia, S.K. Verma. *Prostaglandins Leukot Essent fatty Acids*, 1995, **52**, 223.
18. J.L. Witztum, D. Steinberg, *J. Clin. Invest.*, 1991, **88**, 1785.
19. M. Aviram, E.L. Bierman, A. Chait, J. Biol. Chem. 1988, 25, 15416.
20. S.M. Grundy, S.M. Dinke, *J. Lipid. Res.* 1990, **31**, 1149.
21. H.N. Ginsberg, *Med. Clin. North. Am.* 1994, **78**, 1.

DIETARY FLAVONOIDS PROTECT LOW DENSITY LIPOPROTEIN FROM OXIDATIVE MODIFICATION

G.T.McAnlis,[1,2] J.McEneny,[1] J.Pearce[1] and I.S.Young[2]

[1]Department of Food Science, The Queens University of Belfast, Newforge Lane, Belfast, BT9 5PX.
[2]Department of Clinical Biochemistry, The Queens University of Belfast, Royal Victoria Hospital, Belfast, BTI2 6BJ

1 INTRODUCTION

It is well known that diets rich in fruit and vegetables are protective against cardiovascular disease and it has been suggested the flavonoids, a group of polyphenolic compounds found widely distributed throughout the plant kingdom, may have a role to play in their protective effect. Over 4000 different flavonoids have been identified. Flavonoids are classified according to position and degree of carbonyl and hydroxyl group substitutions on the A, B and C rings of the molecule. The major flavonoid classes include, flavonols, flavones, flavanones found in fruit and vegetables and catechins found in tea.

The daily Western intake of flavonoids has been estimated in the range of 0.5-1.0 g.[1] More recently however it has been suggested that the actual intake is lower than this, around 23mg flavonol (quercetin, kaempferol and myricetin) and flavone (luteolin and apigenin) aglycones.[2] The main sources of dietary flavonoids are thought to be tea, onions, apples and red wine.[2]

Interest in the area of flavonoids and coronary heart disease (CHD) has been stimulated by the discovery that the French have a low incidence of CHD despite having a high fat diet and high prevalence of smoking (the 'French Paradox'). One factor which may contribute to this is the high consumption of red wine in France, an idea which was strengthened in 1993 when it was shown that polyphenols in red wine could inhibit the oxidation of LDL,[3] a step thought to be important in the pathogenesis of atherosclerosis.[4]

A number of epidemiological studies published in the last five years have suggested flavonoid intake may be linked to the incidence of CHD. The Zutphen elderly study,[5] the Seven Countries Study,[6] the Finnish cohort study[7] and most recently, the Zutphen Study[8] all showed a negative correlation between flavonoid intake and mortality from CHD. The epidemiological evidence is, however, not completely clear as the recent publication of a study involving American male health workers found no correlation between flavonoid intake and CHD.[9]

The mechanisms by which flavonoids inhibit the oxidative modification of LDL are unclear. Flavonoids may act as free radical scavengers, metal chelators or both, or they may protect the endogenous vitamin E from oxidation.[10] The aim of this study was to investigate the free radical scavenging and metal chelation ability of the main dietary flavonoids at various concentrations on the oxidative modification of LDL, and to see the effect flavonoids may have on vitamin E oxidation.

2 MATERIALS AND METHODS

2.1 LDL Isolation and Oxidation

Plasma was obtained from heparinised blood taken from a healthy non-smoking male volunteer. LDL was isolated from plasma by rapid ultracentrifugation and oxidised as described by McDowell *et al.*[11] The assay used is based on the production of conjugated dienes, which absorb maximally at 234nm. The lag phase prior to the onset of rapid LDL oxidation is seen as a slope with a slow increase in absorbance (Figure 1). The duration of this phase is indicative of the resistance to oxidation of the LDL being examined. Oxidation was carried out in the presence of $2\mu M$ copper or 5mM AAPH, an azo compound which undergoes thermal decomposition at 37^0C and promotes non-transition metal dependent oxidation.

Various flavonoids were added to the LDL including quercetin, kaempferol, myricetin, apigenin, and rutin (a quercetin glycoside). The concentrations were as low as possible to gain an understanding of the protective effect at close to physiological concentrations.

2.2 Measurement of α-Tocopherol in LDL

α-Tocoperol was extracted from the LDL using heptane and measured on high pressure liquid chromatography (HPLC) using UV detection at 292nm under the following conditions: mobile phase 45% methanol, 45% acetonitrile and 10% dichloromethane; flow rate 1.8 ml/min; C18 ODS column.

3 RESULTS AND DISCUSSION

All the flavonoids investigated demonstrated a dose dependent antioxidant effect on copper mediated oxidation of LDL; this is illustrated by the effect of quercetin (Figure 1). Quercetin was the most potent inhibitor of copper-mediated LDL oxidation, followed by rutin, myricetin, kaempferol and apigenin. Quercetin, rutin, myricetin and kaempferol also demonstrated a dose dependent antioxidant effect with AAPH mediated oxidation but they were not as effective as in the copper mediated oxidation (Figure 2). The addition of 5-0.1µmol/L apigenin to LDL resulted in a reduced lag time and increased rate of propagation in response to AAPH oxidation, suggesting a pro-oxidant effect of this flavonoid. Our initial findings suggest the main dietary flavonoids have no protective effect on the oxidation of endogenous α-tocopherol present in the LDL.

The inhibition of LDL oxidation is influenced by a number of structural features of the flavonoid.[10] The plain three-ring flavone nucleus is a poor inhibitor of LDL oxidation, therefore the presence of hydroxyl and carbonyl substitutions is the main influence on antioxidant activity. In general free radical scavenging activity increases as the degree of hydroxyl groups increases with the 3 and 4 positions on the B-ring and the 5 and 7 positions on the A-ring being the most important. The combination of a carbonyl and a hydroxyl group on the C-ring is also important, giving the molecule the ability to chelate metal ions. Quercetin, the most active antioxidant has hydroxyl groups on the 3 and 4 positions of the B-ring and on the 5 and 7 positions on the A-ring making it a good free radical scavenger. Quercetin also has a carbonyl and a hydroxyl group on the C-ring giving it the ability to chelate metal ions. Conversely, apigenin, the weakest antioxidant, is missing the hydroxyl group on the 3 position on the B-ring making it a weaker free radical scavenger. It is also missing the hydroxyl group on the C-ring making it a poor metal chelator.

4 CONCLUSION

In vitro experiments show that flavonoids inhibit LDL oxidation and suggest that the consumption of flavonoid rich foods may have a protective effect against **CHD.** *In vivo* evidence is, however, still required to demonstrate that flavonoids from the diet are absorbed in large enough quantities to protect our LDL from oxidation.

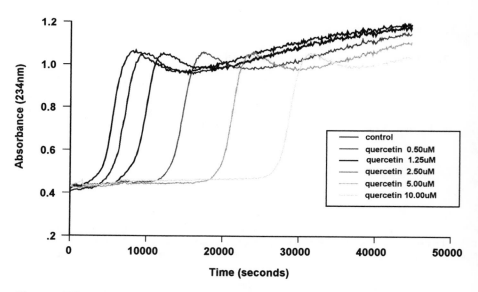

Figure 1 *Effect of increasing concentrations of quercetin on copper induced oxidation of LDL*

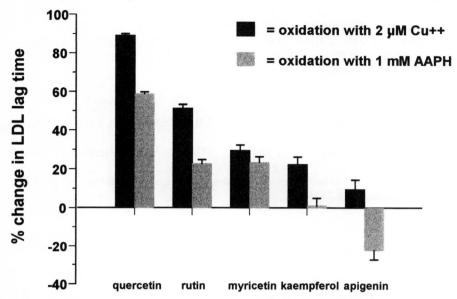

Figure 2 *Effect of 0.5 μmol flavonoids/L on copper and AAPH induced oxidation of LDL*

5 REFERENCES

1. J. Kuhnau, *World review of Nutrition and Diet* 1976, **24**, 117-191
2. M.G.L. Hertog, P.C.H. Hollaan, M.B. Katan & D.Kronihout, *Nutrition and Cancer* 1993, **20**, 21-29
3. E.N. Frankel, J. Kanner, J.B. German, E. Parks & J.E. Kinsella, *Lancet* 1993, **341**, 454-457
4. D. Steinberg, S. Parthasarathy, T.E. Carew, J.C. Khoo & J.L.Witztum, *New Eng. J. Med.* 1989, **320**, 915-924
5. M.G.L. Hertog, E.J.M. Feskens, P.C.H. Hollaan, M.B. Katan & D. Kronlliout, *Lancet* 1993, **342**, 1007-1011
6. M.G.L. Hertog, D.Kronihout, C. Aravanis, H. Blackburn, R. Buzina, F. Fidanza, S. Giampaoli, A. Jansen, A. Menotti, S. Nedeljkovic, M. Pekkarinen, B.S. Simic, H. Tosliima, E.J.M. Feskens, P.C.H. Hollman & M.B. Katan, *Archives o Internal Medicine 1995*, **155**, 381-386
7. S.O. Keli, M.G.L. Hertog, E.J.M. Feskens & D. Kromhout, *Archives of Internal Medicine* 1996, **156**, 637-642
8. P. Knekt, R. Jarvinen, A. Reunanen & J. Maatela, *BMJ* 1196, **312**, 478-4819.
9. E.B. Rmnn, M.B. Katan, A. Ascerio, M.J. Stampfer & W.C. Willet, *Annals of Internal Medicine* 1996, **125**, 384-389
10. C. De-Whalley, S.M. Rankin, J.R.S. Hoult, W. Jessup & D.S. Leake, *Biochemical Pharmacology* 1990, **39**, 1743-1750
11. I.F.W. McDowell, J. McEneny & E. Trimble, *R.Clinical Biochemistry* 1995, **32**, 167-174

ISOFLAVONE PHYTOESTROGENS AS FUNCTIONAL INGREDIENTS IN SOYA PRODUCTS: PHYSIOLOGOGICAL IMPORTANCE OF METABOLISM AND POSSIBLE INDICATIONS FOR THEIR RAPID ABSORPTION FOLLOWING CONSUMPTION OF TEXTURED VEGETABLE PROTEIN

E.A. Bowey,[1] I.R. Rowland,[1] H. Adlercreutz,[2] T.A.B. Sanders[3] and H. Wiseman[3]

[1]Department of Microbiology and Nutrition, BIBRA International, Carshalton, Surrey SM5 4DS.
[2]Department of Clinical Chemistry, University of Helsinki, Meilahti Hospital, FIN-00290 Helsinki, Finland
[3]Nutrition, Food and Health Research Centre, King's College London, Campden Hill Road, London, W8 7AH

1 INTRODUCTION

Dietary phytoestrogens in disease prevention is an area attracting great interest.[1-2] Isoflavone phytoestrogens include genistein and daidzein and their glycoside conjugates genistin and daidzin. Populations in the Far East have been consuming soya beans in the form of traditional products for centuries. Western cultures have only started to adopt soya foods much more recently and Western style soya foods are produced by modern processing techniques in large soyabean-processing plants. In soyabeans and textured vegetable protein (TVP) there are high levels of the conjugated isoflavones. In contrast, in the traditional fermented soyabean products such as miso, nearly all the isoflavones are present in their unconjugated forms.

Isoflavones may exert their effects through their weak oestrogenic properties and other actions, including antioxidant action.[2] Great importance is thus attached to understanding the metabolism of these compounds. When ingested, genistin and daidzin are hydrolysed in the large intestine by bacterial β-glucosidase to genistein and daidzein (biologically active forms), which are then absorbed. Further bacterial metabolism of daidzein by bacterial cytochrome P-450 can produce either the oestrogenic metabolite equol (isoflavan metabolite of the isoflavone daidzein) or O-desmethylangolensin (O-Dma), whereas genistein is probably metabolised to p-ethylphenol. However, we report that in a pilot study carried out to provide preliminary information, consumption of 75g of TVP (containing approximately 55mg of isoflavones) by a female volunteer resulted in increased levels of genistein and daidzein in the plasma only 15 minutes after ingestion. Possible explanations for the observed results are discussed.

2 IMPORTANCE OF ISOFLAVONE METABOLISM

Populations consuming a Western-type diet usually excrete low levels of isoflavonoids in the urine. Urinary excretion of isoflavonoids can be used as a measure of intake and thus possible exposure. In a study comparing the plasma isoflavonoid levels in Japanese and Finnish men consuming their habitual diets, the mean levels of daidzein, genistein, O-Dma and equol were significantly higher in the Japanese men,[3] probably reflecting the low-fat Japanese diet which is rich in soya products.

In a study of postmenopausal Australian women consuming 45g soya flour/day (containing approximately 27mg isoflavones) for 14 days, the mean plasma levels of daidzein and equol were 270nmol/L and 130nmol/L.[4] The influence of chronic soya consumption on the kinetics and urinary excretion of genistein and daidzein has been studied in six young premenopausal women (22-29 years of age) consuming soya milk (containing approximately 34mg daidzein and 39mg genistein, 85% of which were as the respective glucosides daidzin and genistin), with each of three meals daily for one month (total daily intakes were approximately 100mg genistein (as genistin) and 100mg daidzein (mostly as daidzin).[5] At two-week intervals the subjects consumed the full daily allowance of approximately 200mg isoflavones within 30 minutes each day for three to four days (in addition to a constant basal diet) and collected urine continuously.[5] Urinary recovery of genistein and daidzein was initially 24% and 66% of the ingested levels respectively and that of equol was initially 28% of the ingested daidzein + daidzin in one subject and <1% in five subjects. Urinary recovery decreased progressively over the four weeks of daily soya ingestion by 42% for genistein and 31% for daidzein but increased by three to 100-fold for equol in four subjects.[5]

The absorption half-lives for genistein and daidzein decreased, suggesting more rapid absorption. The excretion half-lives for genistein and daidzein decreased suggesting a more rapid excretion. For equol the general trend was towards an increase in half-life suggesting less rapid excretion.[5] The results suggest that chronic soya ingestion can influence the metabolism of ingested isoflavones. In particular increased production of the longer acting more oestrogenic daidzein metabolite equol may have important implications for the overall oestrogenic potency of dietary soya isoflavones.[5] When the analogous experiment was performed in males (six males 21-35 years of age) the urinary recovery of ingested isoflavones did not change with chronic soya ingestion.[6] Alterations in the time courses of excretion were observed. Overall, chronic soya exposure did not alter the pathways of isoflavone metabolism but increased the excretion time courses suggesting chronic exposure to soya could prolong tissue exposure to the unconjugated forms of these isoflavones thus enhancing oncoprotective potential.[6]

The bioavailability of soybean isoflavones has been studied in 12 young adult women at total isoflavone levels of 0.7, 1.3 and 2.0mg isoflavones/kg body weight in soya milk.[7] Urinary recovery of daidzein was greater than that of genistein (21% compared to 9% at all three doses), total fecal excretion of isoflavones was around 1-2% of the ingested amount. The plasma concentrations of daidzein and genistein were approximately equal and total plasma isoflavone concentration was raised to approximately 4400nmol/L at 6.5 hours after the 2.0mg/kg dose. However, urine and plasma concentrations were approximately zero 24 hours after dosing. Daidzein from soya milk appears to be more bioavailable than genistein in adult women.[7]

In a further study the bioavailability of soya isoflavones has been shown to depend on the gut microflora in the women.[8] In seven adult women at total isoflavone levels of 0.89, 1.8 and 2.7mg isoflavones/kg body weight, consumed three times/day (2.67, 5.4 and 8.1mg isoflavones/kg body weight/day), five of the seven subjects excreted only 0.4-0.7% of ingested isoflavones intact in feces, in contrast to 6-8% of ingested isoflavones in the feces of the other two subjects. This difference in fecal excretion greatly influenced isoflavone bioavailability. Urinary recovery of ingested isoflavones was more than twice as high in high verses low fecal excretors.

Gut bacterial enzymes such as β-glucuronidases can hydrolyse isoflavone glucuronide conjugates to aglycones (rapidly absorbed). However, intestinal microflora can also extensively metabolise and degrade isoflavones thus preventing their reabsorption from the lower bowel. If an individual possesses bacteria that are not effective in isoflavone metabolism and degradation then more isoflavones would be absorbed. This might then explain the positive association between high fecal isoflavones and greater total urinary recovery of isoflavones. Urinary excretion of phytoestrogens as a marker for exposure may also provide some indication of cancer risk. A very low urinary excretion of equol in postmenopausal breast cancer patients compared to vegetarians has been observed.[1]

3 ABSORPTION AND METABOLISM OF ISAFLAVONES FROM TEXTURED VEGETABLE (SOYA) PROTEIN

A pilot study was carried out to provide preliminary data on isoflavonoid absorption and metabolism and involved the consumption of 75g textured vegetable (soya) protein (TVP) (containing approximately 55mg isoflavones) by a female adult premenopausal volunteer. The results in Table 1 show the appearance of increased levels of genistein and daidzein in the plasma (34.3nmol/L and 22.4nmol/L, respectively) only 15 minutes after ingestion (compared to the levels at zero time: 5.3nmol/L for both isoflavones). The plasma levels for both isoflavones peaked at six hours (694.37nmol/L and 331.27nmol/L respectively).

This rapid appearance of daidzein and genistein in the plasma as measured by GC-MS, may represent absorption of unconjugated isoflavones from the TVP. Alternatively the results may also suggest either rapid hydrolysis of conjugates in the upper alimentary tract or absorption of intact glycosides and we are exploring these possibilities further.

Further studies are in progress to provide more detailed information concerning both inter-individual variation in isoflavonoid metabolism and to elucidate possible mechanisms of absorption. In particular, we aim to investigate whether differences in isoflavone metabolism are related to differences in the metabolic activity of the gut flora or to other factors related to gastrointestinal function such as fecal moisture, transit time, fecal pH or subjects' habitual diets.

Table 1 *Isoflavonoid Concentration in the Plasma After Ingestion of 75g TVP*

Time (min)	Daidzein (nmol/L)	O-Desmethyl-angolensin (nmol/L)	Equol (nmol/L)	Genistein (nmol/L)
0	5.35	0.00	0.12	5.31
15	22.46	0.49	0.00	34.30
30	69.59	0.78	0.01	116.95
60	130.27	1.06	0.00	271.60
180	206.18	0.83	0.00	535.11
360	331.27	6.00	0.58	694.37
480	318.00	17.43	0.00	605.95

4 ACKNOWLEDGEMENT

This project is funded by EC (FAIR-CT-95-0894).

5 REFERENCES

1. C.H.T. Adlercreutz, B.R. Goldin, S.L. Gorbach, K.A.V. Hockerstedt, S. Watanabe, E.K. Hamalainen, M.H. Markkanen, T.H. Makela, K.T. Wahala, T.A. Hase and T. Fotsis *J. Nutr.* 1995, **125** 757S2.
2. H. Wiseman *Biochem.Soc.Trans* 1996, **24** 795-8003.
3. H. Adlercreutz, H. Markkanen and S. Watanabe *Lancet*, 1993, **342** 1209-12104.
4. M.S. Morton, G. Wilcox, M.L. Wahlqvist, and K. Griffiths *J.Endocrinol.* 1994, **142** 2515.
5. L-J.W. Lu, S-N. Lin, J.J. Grady, M. Nagamani and K.E. Anderson, *Nutr.Cancer* 1996, **26** 2896.
6. L-J. W. Lu, S-N. Lin, J.J. Grady, M.V. Marshall, V.M. Sadagopa, and K.E. Anderson *Nutr.Cancer* 1995, **24** 3117.
7. X. Xu., H-J. Wang., P.A. Murphy, L. Cook and S. Hendrich *J.Nutr.* 1994, **124** 8258.
8. X. Xu., K.S. Harris, H-J. Wang, P.A. Murphy and S. Hendrich *J.Nutr* 1995, **125** 2307

DIALYSABILITY OF CALCIUM, MAGNESIUM AND IRON AS A METHOD TO COMPARE *IN VITRO* BIOAVAILABILITY OF MINERALS IN FOODS

C. van Aarle, E. Bontenbal and R.E. Potjewijd

PURAC biochem
PO Box 21
4200 AA Gorinchem
The Netherlands

1 INTRODUCTION

The aim of this study was to investigate the main factors influencing the *in vitro* bioavailability of calcium, magnesium and iron. Different mineral salts were tested in common foods. The method used is based on a simulated gastric digestion followed by a measurement of soluble or dialysable mineral, under physiological conditions.

1.1 Bioavailability

Bioavailability has been defined as '*the proportion of a nutrient capable of being absorbed and available for use or storage*,' more briefly '*the proportion of a nutrient that can be utilised*'.[1]

Bioavailability can be measured in two ways: with *in vitro* methods and *in vivo* methods. The *in vitro* methods are done outside an organism, but in laboratory equipment. One *in vitro* method is based on the determination of the amount of the dialysable component under simulated physiological conditions. The advantages of *in vitro* measurements are that they are cheaper, allow a large sample capacity, give accurate and reproducible results and are not limited by ethical constraints. The limitations are, however, that no absolute absorption can be measured and that only the passive absorption is measured.[2]

Factors controlling the bioavailability of minerals can be divided in endogenous and exogenous factors. Endogenous factors are: age, sex, pregnancy and lactation. Exogenous factors include a number of dietary variables that may influence the absorption such as vitamins, fibre, phytase, oxalate, fat and lactose, and a number of dietary variables that may influence urinary excretion.[3]

1.2 Minerals

1.2.1 Calcium. Calcium is the most abundant mineral in the body, being about 2% of body weight. Most of the calcium (99%) is found in the skeleton and in the teeth.[4] Calcium plays a role in blood coagulation, blood pressure, contraction of muscles and bone density. An individual's bone mass is primarily genetically determined. However, dietary intake and physical activity have also an influence.[5] A low bone mass can be the result of bone loss during aging, or a failure to achieve sufficient peak bone mass in early adulthood.[6] This low bone mass is characteristic for osteoporosis, which increases the susceptability to fractures.

The major sources of calcium in the western diet are dairy products, which contribute 35-67% of the calcium intake.[7] On average about 30% of the calcium ingested is absorbed, but this is affected by several dietary and physiological factors.

Dietary factors which affect calcium bioavailability include the chemical form and the solubility of the calcium, as well as the presence of inhibitors in the food. Substances which form insoluble complexes with calcium in the intestines, such as phytate, oxalate and uronic acid and some non starch polysaccharides (NSP), including fibres, have all been reported to reduce calcium bioavailability. High levels of protein in the diet increase the absorption, but also lead to an increased urinary excretion, so that the net result is a decrease in calcium bioavailability.[8]

1.2.2 Magnesium. The human body contains about 20-28g magnesium, of which 55% is stored in the skeleton. Magnesium plays an important role in the breakdown and synthesis of carbohydrate, fat and protein and in the mechanism of muscle relaxation.

The kidneys play a dominant role in the excretion of magnesium. Since calcium and magnesium partly use the same transport sites in the renal tubuli, reabsorption of magnesium is inversely related to the reabsorption of calcium.[3] The intestinal absorption varies between 35-70% depending on the intake levels.[9]

Phosphate, calcium, alcohol, oxalate, dietary fibres and free fatty acids decrease magnesium absorption, while protein, lactose and vitamins D and B_6 increase absorption.[10]

1.2.3. Iron. The body contains about 3-4g iron. Two-thirds of the total iron is stored in the haem of red blood cells. Five to ten percent is used in myoglobin, cytochrome enzymes and other enzymes. In bone marrow, liver and spleen, iron is stored in ferritin and haemosiderin form.[11] Iron deficiency is a major nutritional problem in the world. The high prevalence of iron deficiency in most developing countries is due to the lack of haem iron in the diet.[12]

Iron is absorbed by an active saturable process, primarily in the duodenum.[13] The body regulates iron homeostasis by controlling absorption and not by modifying the excretion as with most minerals. Haem and nonhaem forms of iron are absorbed by different mechanisms. Haem iron is highly absorbable. The absorption of nonhaem iron is strongly decreased by calcium, phophate, phytates, polyphenols. The availability of iron is enhanced by ascorbic acid.[4] The percentage of iron absorbed from a meal decreases as the amount of iron present increases.[14]

2 METHODS

The method used was based on a simulated gastric digestion followed by measurement of soluble or dialysable mineral, under physiological conditions.[15,16] This system simulates the peptic digestion in the stomach, the gradual pH change in the duodenum and the pancreatic digestion followed by absorption. The mineral dialysability was measured after one and three hours pancreatic digestion and the average was taken. The amount of mineral was determined in both the retentate and the dialysate by Flame Atomic Absorption Spectrometry (FAAS).

The various foods which were fortified were wheat bran, baby food, orange juice, diet meal replacement, breakfast cereal, energy drink and baby formula. The foods were fortified with 300mg calcium, 100mg magnesium and 15mg iron per standard portion.

3 RESULTS AND DISCUSSION

The results showed that both the mineral source and the food matrix have an influence on the amount of mineral dialysed.

3.1 Influence of the Mineral Source

Taken over all the different foods, the amount of dialysed mineral ranged between 85% and 66% for calcium, between 90% and 84% for magnesium and between 39% and 32% for iron, depending on the mineral source.

Independently of the food matrix, soluble calcium salts were better dialysed than insoluble salts or salts soluble only at low pH. Calcium lactate demonstrated the highest amount dialysed (85%), followed by citrate (80%). Calcium phosphate was the least dialysed (66%). This is in line with the difference in solubility between the calcium salts.

The magnesium source did not significantly influence the amount dialysed. All magnesium salts were well dialysed and therefore highly available for absorption. Citrate was the most dialysed (90%) and gluconate the least (84%).

The amount of iron dialysed was low (35%) for all iron sources. Ferrous lactate was the most dialysed (39%), followed by gluconate (34%). The least soluble iron sources were also the least dialysed.

3.2 Influence of the Food Matrix

The results demonstrated that the food matrix influences the amount of mineral dialysed to a similar extent to the different mineral sources.

The dialysability of calcium was negatively influenced by fibre, where the average calcium dialysed was only 54% in wheat bran. In the presence of fibre, calcium phosphate was the least dialysed (34%), followed by calcium carbonate (49%). Calcium citrate was the most dialysed (67%) followed by lactate (63%). This effect could be explained by the size of the molecules citrate and lactate, which could protect the calcium ion from being attracted and bound by fibre. The other food matrixes did not significantly influence the calcium dialysability.

The various food matrixes did not have a significant influence on the magnesium amount dialysed.

The dialysability of iron was the most affected by the food composition. The highest amount dialysed was in orange juice (78%), the lowest in wheat bran (12%). This can be explained by the higher amount of iron in the ferrous form at a low pH and by the attraction of the fibre which results in insoluble complexes. The negative effect of fibre was stronger with ferrous sulphate than with ferrous lactate. An explanation could be the protective effect of the bigger organic molecule.

4 CONCLUSIONS

The followed method allowed a good comparison between mineral salts and an evaluation of the importance of the food composition. The results gave an indication of the amount of each mineral which is available for absorption. It does not however assess the interactions between the minerals.

5 REFERENCES

1. A. E. Bender,'Nutrient availability: Chemical and biological aspects', Southgate, The Royal Society of Chemistry, Cambridge, 1989, p.3-9.
2. R. Havenaar and M. Minekus, *Dairy Ind Intl*, 1996, **9**, 17.
3. G. Schaafsma, *Eur J Clin Nutr*, 1997, **51**, 15.
4. National Research Council, 'Recommended Dietary Allowances', National Academy Press, Washington DC., 1989, 10th edition, p. 187.
5. D. Teegarden and C.M. Weaver, *Nutr Rev*, 1994, **52**, 171.
6. D.C. Welten, 'Calcium intake in relation to bone health during youth', Thesis Publishers, Amsterdam, 1996.
7. G.S Ranhotra, *Cereal Foods World*, 1986, **31**, 535.
8. A. Halliday, *BNF Nutrition Bulletin*, 1991, **16**, 11.
9. M.E. Shils, 'Present Knowledge in Nutrition', The Nutrition Foundation, Washington, DC., p. 423.
10. E.J. Brink, P.R Dekker and E.C.H. van Beresteijn, *Voedingsmiddelentechnologie*, 1989, **13**, 22.
11. J.F. Wijn and W.A. Staveren, 'De Voeding van elke dag', Bohn, Scheltema & Holkema, Utrecht, 1986.
12. D. James and M.D. Cook, *Food Technology*, 1983, 124.
13. R.W. Charlton and T.H. Bothwell, *Ann Rev Med*, 1983, **34**, 55.
14. R.F. Hurrell, *Eur J Clin Nutr*, 1997, **1**, 1.
15. D.D. Miller, B.R. Schricker, R.R.Rasmussen and D. Van Campen, *Am J Clin Nutr*, 1981, **34**, 2248.
16. T. Hazell and I.T. Johnson, *Br J Nutr*, 1987, **57**, 223.

FUNCTIONAL ANTIOXIDANT PROPERTIES OF A NOVEL FOOD COLOURANT PRODUCED FROM AN APPLE-DERIVED FLAVONOID USING POLYPHENOL OXIDASE FROM AN APPLE SOURCE

T.J. Ridgway,[1] J.D. O'Reilly,[2] G.A. Tucker,[1] and H. Wiseman[2]

[1]Department of Applied Biochemistry and Food Science, University of Nottingham, Sutton Bonington Campus, Loughborough, Leicestershire, LE12 5RD
[2]Nutrition, Food and Health Research Centre, King's College London, Campden Hill Road, London, W8 7AH

1 INTRODUCTION

In recent years there has been a market led removal of synthetic additives from food produce; in particular, the removal of colourants, for example the yellow azo dye tartrazine (E 102), from soft drinks and squashes.[1] Likewise there has been an increasing interest in the health promoting properties of secondary plant metabolites, especially as mediated by the antioxidant properties of the coloured carotenoids and anthocyanins.[2] Natural colourants, including anthocyanins and carotenoids, would appear therefore, to represent ideal synthetic colourant replacements; not only could they provide a technical colour function, but they could also help to maintain the health of the consumer. The relative instability of natural colourants compared with synthetic colourants, however, is a considerable obstacle to their more widespread use.[3]

Here, the colourant and antioxidant properties of the phloridzin (flavonoid) derivatives phloridzin (dimerised) oxidation product (POP) and its aglycone phloretin (dimerised) oxidation product (PeOP) are described, together with the antioxidant properties of the additional phloridzin derivatives, 3-hydroxyphloridzin and 3-hydroxyphloretin. The results suggest that blends of phloridzin derivatives may be used as a natural alternative to tartrazine, combining excellent technical performance (including high stability in soft drinks) with potentially health beneficial antioxidant properties. The inclusion of the precursors phloridzin and phloretin in blends, could, in addition, provide potentially health beneficial weak oestrogenic effects.[4]

2 THE PRODUCTION OF PHLORIDZIN DERIVATIVES

Phloridzin is a dihydrochalcone largely restricted to apple (*Malus sp.*). Although phloridzin is only found in limited quantities in the mature apple fruit, in young apple leaves and twigs phloridzin accounts for up to 10% of the dry weight. To enable the use of phloridzin as a precursor for food additive production, purification methods suitable for large scale application have been developed.[5,6] A water based extraction method was developed to reduce costs, but the actual method utilised would depend on a number of factors including the value of phloridzin derivatives. To aid phloridzin production, it would appear preferable to grow apple plants as coppiced rootstocks, vigorous rootstock types such as M25 appearing to be particularly productive.[5,6]

Potentially economic methods have been developed for the purification of apple polyphenol oxidase (aPPO) using standard chromatographic media, including, DEAE-sephadex, phenyl-sepharose and S-200.[7] The oxidation products of phloridzin were produced using aPPO, as described below. In contrast to fungal tyrosinase, aPPO is particularly effective at catalysing the 3-hydroxylation (by EC 1.14.18.1. activity) of phloridzin.[8] In common, though, with most polyphenoloxidases, aPPO also shows good *o*-diphenol oxidase activity and hence 3-hydroxyphloridzin produced by the oxidation of phloridzin is immediately converted to a quinonic form. Subsequently shifting the pH to pH 8 (from the enzyme optimum of pH 5), gives two isomeric coloured dimers, the structures of which have not yet been fully characterised.[7]

L-Ascorbic acid can be used to partially block the formation of quinones; it achieves this by continually reducing them to the *o*-diphenol form as they are formed.[7,8] The use of L-ascorbic acid thus allowed the production of 3-hydroxyphloridzin. The aglycone forms of phloridzin and 3-hydroxyphloridzin were produced by acid hydrolysis. Oxidation by aPPO and dimerisation of phloretin in 20% ethanol allowed the formation of PeOP.[8] The aglycone forms offered increased lipophilicity and thus extended the potential range of phloridzin derivative application.

3 COLOUR PROPERTIES OF PHLORIDZIN DERIVATIVES (POP and PeOP)

At the pH of soft drinks (typically pH 3 to 4) the appearance of POP resembles that of the synthetic colourant tartrazine and the natural colourant curcumin. Using Hunter Lab apparatus and transmitance through 1cm path cuvettes, the maximum degree of greenness (negative value on Hunter Lab, **a** axis) of POP was observed to be -18, which compared with -16 for curcumin and -14 for tartrazine. Though POP has a 'sharp' and 'crystal clear' appearance (almost fluorescent) for some applications the degree of greenness may prove unsuitable. PeOP, however, was found to show a maximum greenness of only -11 such that blends of POP and PeOP could potentially span all the shades shown by the currently used lime-yellow colourants; indeed most additive-coloured foods contain a blend of colourants to give the desired shade.

A 0.001% solution of tartrazine was found to give an absorbance of 0.4 AU at 420nm (λ max.). POP at soft drinks pH 3 to 4, showed a slightly greater absorbance; a 0.001% solution giving an absorbance of 0.5 at 410nm (λ max.).[8] Values of pH above pH 5.5, gave POP an orange appearance. The absorbance given by a 0.001% solution at pH 8 was 1.0 AU at 470nm (λ max.).[8] Potential products which could utilise higher pH values (pH 5 to 7) could include novel 'iced-tea' drinks; part of the rising market for new adult soft drinks.[9]

In preliminary colour stability trials POP and PeOP proved to be more stable than tartrazine and curcumin under many conditions; for example, in fluorescent light at constant 22°C in commercial produce (lemonade and cider).[8] It appeared therefore that POP and PeOP confirmed the predictions of Taylor,[3] that dimerised flavonoid oxidation products could be expected to be relatively stable chromophores. Importantly POP and PeOP are the products of oxidative bioconversion. This is in contrast to the oxidation of unsaturated hydrocarbon chains, as found in curcumin or carotenes, which causes reduced colouration through the loss of conjugation sequences.

4 ANTIOXIDANT PROPERTIES OF PHLORIDZIN DERIVATIVES

Using physiologically relevant model membrane antioxidant assays, the antioxidant properties of PeOP, POP and other phloridzin derivatives were compared with the antioxidant properties of quercetin in Table 1. Quercetin is the dietary flavonoid usually found to show the greatest lipophase antioxidant ability.

The coloured phloridzin derivative PeOP, was shown to have good antioxidant properties, though its glycone form POP showed no antioxidant activity in the lipophase test systems used. The phloridzin derivative 3-hydroxyphloretin showed excellent antioxidant activity, in the microsomal system exceeding that of quercetin.

The structure-function relationship of flavonoids in relation to antioxidant activity is well characterised,[9,10] the principle features being the number of hydroxyl groups and the possession of *o*-diphenolic structures which readily take part in redox coupled reactions. Glycone forms show lower antioxidant activity in lipid based test assays than aglycones due to their relative inability to access the damaging free radicals. 3-Hydroxyphloretin which has five free hydroxyl groups, two of which form an *o*-diphenol structure, could thus have been predicted to show good antioxidant activity, and had previously been shown to be an excellent antioxidant in lard.[11]

Table 1 *Antioxidant Properties of Phloridzin-derived Aglycone Colourant*

Compound	Liposomal IC_{50} (μM) Fe/Asc	Systems Microsomal IC_{50} (μM) Fe/Asc	Microsomal IC_{50} (μM) Fe-ADP/NADPH
PeOP	7.2	6.0	22
POP	NR	NR	NR
3-Hydroxyphloretin	2.8	1.4	2.2
3-Hydroxyphloretin	6.8	17	19
Phloretin	27	14	11
Phloridzin	NR	NR	NR
Quercetin	1.5	1.6	4.3

NR: 50% inhibition not reached

IC_{50} values are shown for the inhibition of liposomal and microsomal lipid peroxidation.

Values were determined from graphs of concentration-dependent inhibition of Fe/Ascorbate or Fe-ADP/NADPH mediated lipid peroxidation in which each point on the graph represented the mean $\pm$ SD of six separate assays (data not shown).

5 FUNCTIONAL BENEFITS/TOXICOLOGY OF PHLORIDZIN DERIVATIVES

POP, phloridzin and phloretin can be shown to be part of the existing diet, a fact which should aid obtaining legislative permission for use of coloured and antioxidant phloridzin derivatives. POP forms a large proportion of the colour of apple juices and ciders,[12] and phloridzin and phloretin are found in fresh apple fruit.[8] A typical cored, but not peeled, desert apple fruit, contains 3mg phloridzin.[8] Uptake of phloridzin in the gut has been demonstrated with blood plasma levels of 1.6 µM having been reported in an individual on a 'normal diet'.[13]

The potential health benefits of flavonoids and isoflavonoids are increasingly appreciated,[10] including the beneficial effects of antioxidant and weak oestrogenic properties. PeOP, 3-hydroxyphloretin and 3-hydroxyphloridzin have been described here (Table 1) as effective antioxidants and phloretin has been shown to have weak oestrogenic properties.[14] Both properties are of potential importance in the prevention of coronary heart disease and cancer, but both cannot be optimised on the same molecule.[4,6]

In addition phloridzin and phloretin affect facilitated glucose uptake.[15] This could be regarded as an undesirable toxic effect, but may be of value in the Western diet at dietary levels, in the prevention of diabetes, and could possibly be of use in cancer therapy at supra-dietary levels. Dose is likely to be a critical factor in distinguishing between the effects of phloridzin derivatives as, non-functional food additives, functional foods, preventative medicines, therapeutic medicines or, in great excess, poisons. Care evidently needs to be taken in assessing the potential benefits and risks of all non-nutrient micronutrients/functional food additives.

Pending full toxicological assessment, the potential of PeOP as an additive of both technical (colour) and functional (antioxidant) capability, however, is clear. Furthermore, the use of blends of phloridzin derivatives could enable the production of a range of products with, yellow colouring, antioxidant, and weak oestrogenic properties, as required.

6 ACKNOWLEDGEMENTS

This project was funded by Nottingham University and MAFF.

7 REFERENCES

1. M. Knott, Colours, a consumer whitewash? *Food Manufacture*. 1989, **Feb**, 21.
2. R. I-San Lin, in 'Functional foods, pharmafoods, nutraceuticals' (ed., I. Goldberg), 393.Chapman and Hall, New York, 1995.
3. A.J. Taylor, in 'Developments in food colours 2' (ed., J. Walford), 159, Applied

science publishers, London, 1984.

4. T.J. Ridgway and G.A. Tucker, *Biochem. Soc. Trans.*, 1997, **25**, 59.

5. T.J. Ridgway and G.A. Tucker, *Biochem. Soc. Trans.*, 1997, **25**, 110S.

6. T.J. Ridgway and G.A. Tucker, *Biochem. Soc. Trans.*, 1997, **25**, 109S.

7. T.J. Ridgway, G.A. Tucker and H. Wiseman, in 'Biotech. and Genetic Eng. Reviews (ed., M. Tombs) volume 14 . Intercept Press, Andover (in press).

8. T.J. Ridgway, Ph.D. Thesis, University of Nottingham, 1996.

9. W. Bors and M. Saran, *Free Rad. Res. Communs.*, 1987, **2**, 289.

10. H. Wiseman, *J. Nutr. Biochem.*, 1996, **7**, 2.

11. S.Z. Dziedzic, B.J.F. Hudson and G. Barnes, *j. Agric. Food Chem.*, 1985, **33**, 244.

12. A.G.H. Lea, *Journal of Chromatography*, 1982, **238**, 253.

13. G. Paganda and C.A. Rice-Evans, *FEBS Letters*, 1997, **401**, 78.

14. R.J. Miksicek, *Molecular Pharmacology*, 1993, **44**, 78.

15. H.Neef, P.Augustijns, P.Declerq, P.J.Declerk, and G. Lacheman, *Pharmaceutical and Pharmacological Letters*, 1996, **6**, 86.

III Technological Aspects

FUNCTIONAL FOODS: ASSURING QUALITY

A.J. Alldrick

Cereals & Cereal Processing Division
Campden & Chorleywood Food Research Association
Chipping Campden GL55 6LD

1 INTRODUCTION

Improvements in public health have led to a longer lived population. This is amply illustrated by the observation that, in England, life expectancy at birth has risen from 44 and 47 years for males and females respectively in 1840 to 73 and 78 years respectively in 1989.[1] These statistics are complemented by changes in the spectrum of causes of death. Between 1931 and 1991 the proportion of deaths attributable to circulatory diseases and cancer in England and Wales rose from 26% and 13% to 46% and 25% respectively.[2] By way of contrast, deaths due to infectious and genito-urinary diseases fell from 13% and 5% respectively, to a point where they were included in the 'other causes' category for 1991.[2] Similar changes have been seen in causes of premature (before age 65 years) death.

These changes have provoked challenges for society as a whole, and in particular for government, the medical and allied professions, and the food industry. The reasons for this are numerous but include

- the increased cost of caring for those suffering from ailments which, until recently, were rare, incurable and/or unmanageable
- a recognition that many diseases afflicting the population are multifactorial in origin
- a recognition that lifestyle (in its widest sense) is a significant contributory factor
- that one attribute of lifestyle which plays an important part in determining a population's susceptibility to certain diseases is diet.

Nutritionists, in particular, are facing a considerable challenge in countries such as the UK, where a large proportion of the population enjoy a plentiful, diverse and affordable food supply. From dealing with classical nutrition-linked diseases of deficiency (eg starvation and nutrient deficiency diseases), nutritionists are now faced with diseases which are frequently associated to one degree or another with dietary excess. The challenge is therefore one of identifying dietary practices which not only can assist in achieving a healthier lifestyle, but are also acceptable to the general public.

This is an immense task. The development of the consumer society, with all that it entails, means that government intervention in the Nation's diet, for example, by rationing (as last seen during World War II and the years immediately after), or by some form of financial intervention (subsidy or surcharge) is not feasible. The social changes seen in countries such as the UK have other implications on what the population eats and how much it pays for it. One consequence of these social changes has been that the proportion of disposable income spent on household food purchases in the UK has fallen from about 20% in 1970 to less than 12% in 1990.[3] These observations have led to a number of questions. Two particularly important questions are:

- How can a population's diet be altered in a way that is both acceptable to the consumer and contributes to a healthier life style?
- How can market shares and profits be maintained and preferably improved in what is (certainly in the UK) a mature market?

One answer to both of these questions has been functional foods. The Japanese Ministry of Health and Welfare[4] has described functional foods as *'processed foods containing ingredients that aid specific bodily functions in addition to being nutritious.'* In order for something to be described as a functional food, the Ministry required that:

- it would be a genuine food, not in the form of a capsule or powder and would be derived from naturally occurring ingredients
- it could and should be eaten as part of a daily diet
- it had a particular physiological effect when eaten.

This is a broad definition and arguably embraces a wide range of foods ranging from alcoholic beverages, through fibre-rich breakfast cereals to yoghurts containing added microorganisms (bioyoghurts).

In order to restrict the scope of this chapter to manageable proportions, I will define functional foods as *'processed foods containing ingredients that aid bodily functions in addition to being nutritious and for which some form of health claim is made.'* In this context, the term 'health claim' is defined in terms of the draft guidelines issued by the UK Ministry of Agriculture Fisheries & Food (MAFF).[5] Such a claim would be *'any statement, suggestion or implication in food labelling or advertising that a food is beneficial to health, but not including nutrient content claims .... nor medicinal claims.'* It should be remembered that under UK Law, claims that a food can cure, treat or prevent a human disease or any reference to such properties is normally prohibited.[6] For the purposes of this chapter and again using the MAFF guidelines I will exclude foods for particular nutritional uses (PARNUTS) or those with a direct medical function (eg parenteral feeds).

Functional foods therefore are foods which are consciously eaten, not only to satisfy fundamental dietary needs, but also to elicit additional physiological effects (eg improved intestinal function in the case of fibre-rich breakfast cereals and bioyoghurts). In terms of quality therefore, functional foods must not only meet standards to satisfy good manufacturing practice and public safety, but also must demonstrably achieve the additional claimed physiological effects. Quality has been defined as *'the totality of features and characteristics of a product or service that bear on its ability to satisfy stated or implied needs.'*[7] The aim of this chapter is to discuss the processes that need to be in place to assure the quality of functional foods.

2 ASSURING QUALITY

2.1 The Implications of Quality

The definition quoted above makes it clear that the term quality requires consistency in achieving specification. In the case of foods in general, this means that the producer must be able to demonstrate that such foods are produced and supplied efficiently, safely, and of the standard required in terms of ingredients and eventual sensory attributes such that within-product variation is kept to a minimum.

As alluded to above, in the case of functional foods, there is a further requirement, that for a substantial proportion of a given population it is possible to demonstrate that a functional food delivers the desired physiological effect which is the subject of that claim, and that the effect is sustainable. Thus, as in the case of 'ordinary' foods but even more so in the case of functional foods, quality applies not only to the manufacture and delivery of the product but also to its design.

2.2 The Reality of Quality

2.2.1 Product Design. Before any product is marketed, it has to undergo a design stage during which a concept is transformed into something that can be consistently produced and sold. The design process must therefore be subject to control and verification. An example of an appropriate quality system to control design is that described in the standard BS EN ISO 9001 : 1994.[8] Three parts of the section on design control (section 4.4) within the standard stand out clearly as being essential requirements in any design process. These are that:

- design input requirements need to be identified for the product and that these must include statutory and regulatory aspects
- design output must be measured (this includes identification of *'those characteristics of the design that are crucial to the safe and proper handling of the product,'*)
- design validation should be undertaken *'to ensure that the product conforms to defined user needs or requirements.'*

While not every food company seeks ISO 9001 certification, many do seek certification under one of the third party food-quality schemes recognised by certain retailers. One of the requirements of food-quality standards, such as that operated by the European Food Safety Inspection Service (EFSIS, a leading third party auditing service for the food industry) for accreditation is that *'Products should be developed to deliver acceptable levels of quality and safety and fully meet customer and user requirements'.*[9]

This will present a number of challenges to designers of functional foods, since attention will have to be paid not only to identifying the appropriate functional ingredient and the amount that will achieve a statistically significant effect but also that the effect can be seen in the target population. Since the product will often contain pharmacologically active ingredients, designers of functional foods will have to ensure that the product is safe (see also section *2.2.3 Product Safety* below). This will necessitate an understanding of the toxicological qualities of the ingredients, both singly and in combination, and identifying any subpopulations which might be at risk. Examples of such groups include those with specific food allergies or intolerance syndromes, children, pregnant women and the elderly. These considerations will have to be considered not only at the design input/output stages, but also at the design validation step.

A hypothetical example of how this might impact on design input/output would be a functional food which relied on iron for its additional functionality. In this case, it would be necessary to take into account the chemical form in which the iron was added. This is because, depending on its origin, the bioavailability (the amount of a material that is actually assimilated from the food by the consumer) of iron is dependent on its dietary source.[10] Haem and non-haem iron are absorbed from the gut by different processes. Haem iron (found in haemoglobin and myoglobin in animal foods) is relatively available and its absorption is unaffected by factors such as other dietary components and the iron status of the consumer. In contrast the absorption of non-haem iron is highly variable, depending on the composition of the meal and the physiology of the consumer.

Designers of functional foods will also face the challenge of validating the design of their product. This will not only require a demonstration that the food is produced safely and consistently but also that the food has the claimed biological effects apart from those of simple nutrition. This is implicit in terms of either the standard ISO 9001 or food-specific quality standards. It is also a view which has also been expressed in the MAFF draft guidelines.[5] These propose that manufacturers of functional foods substantiate health claims with a dossier of supporting scientific evidence. The dossier should, amongst other things, have evidence of the specific physiological effect, toxicological data and, where applicable, a history of previous use. The draft guidelines go on to require that evidence for the health claim should be from human studies and include epidemiological evidence. This evidence should:

•   be relevant to the UK population as a whole or to the target population
•   be based on reasonable levels of consumption (in terms of both frequency and amount)
•   demonstrate that the benefit is not a short term response to which the body adjusts
•   take into account confounding factors
•   be statistically significant.

These are rigorous requirements which will require attention, when formulating and justifying health claims. For example, it is relatively easier to demonstrate that regular consumption of a product will contribute to a lowering of plasma cholesterol concentration (a risk factor in cardiovascular disease[11]), than that eating the product will reduce one's risk of cardiovascular disease *per se* (which could also raise the accusation that the claim implied that the product prevented the disease, a possible contravention of current UK legislation[6]).

Note the use of the words 'relatively easier.' Epidemiology is essentially a statistical science and is susceptible to controversy. An illustration of the problems that could be faced is exemplified by the claimed plasma cholesterol lowering ability of dietary soluble β-glucans, in particular those found in oats. The literature contains a number of studies which demonstrate, or fail to demonstrate, that consumption of soluble β-glucans has a reductive effect on plasma cholesterol concentrations. In order to obtain some form of consensus, Ripsin *et al.*[12] performed a meta-analysis of these studies. After eliminating those for which there were inadequate controls, 20 trials were included in the final analysis. From this analysis, Ripsin and her colleagues determined that consumption of oat soluble β-glucans was associated with a small but sustainable reduction in plasma cholesterol concentrations. Essentially the higher the subject's initial plasma cholesterol concentration and the greater the amount of soluble β-glucan eaten, the greater the reduction. However it is important to note that these reductions (between 0.09 and 0.41 mmol/litre are actually quite small and for most of the UK population, would, on their own, fail to bring their plasma cholesterol concentrations to a desirable level (<5.2 mmol per litre).[11]

2.2.2 Product Manufacture. Modern food manufacturing practices have altered dramatically in recent times. The trend has been away from simply ensuring that mistakes did not reach the customer and end-user (quality control) to a complete management system which includes quality control and is designed to ensure that the probability of those mistakes occurring is kept to a minimum (quality assurance). Modern quality management systems are essential in the food industry to guarantee delivery to specification (not just in terms of composition, sensory attributes and safety, but also commercial parameters, eg profitability and reliability).

One of the most important areas of quality assurance in the food industry is the maintenance of the integrity of raw materials and of their properties which are desired in the finished product. This is particularly important in the case of functional foods where it would be clearly undesirable to start with raw materials devoid of the ability to elicit the desired effect or of processes (either during manufacture or subsequent storage of the product) which led to a deterioration of that effect. Nor would it be desirable to have processes, which devoid of appropriate monitoring and control systems, could compromise product quality in general and functionality specifically in the event of a malfunction. It is therefore necessary that ingredient suppliers, and perhaps producers of functional foods as well, have positive release systems which ensure that functional ingredients are evaluated and confirmed in their functionality before being used in manufacture. In addition, manufacturing, storage and delivery systems must be of a standard to ensure product quality from point of manufacture to purchase by end user.

Examples of the raw materials which might fall into this category are the probiotic microbial cultures added to certain dairy products (eg bioyoghurts). In designing a manufacturing and distribution system for such a product it will be necessary to:

- maintain the genetic integrity of the probiotic organisms (ie avoid 'genetic drift')
- ensure that they are cultivated in a manner that excludes the possibility of contamination with foreign organisms (see also section *2.2.3 Product Safety, below*)
- ensure that when added to the product they survive long enough to be consumed in sufficient quantities to have the desired physiological effect.

The clinical efficacy of probiotic cultures has been the subject of some debate[13,14] which will not be discussed further here. However, it is important to note that the literature contains the results of surveys of probiotic materials which show that numerous products either failed to have the named organism(s) for which a claim had been made, or had other microorganisms in addition to those claimed.[15,16] These incidents highlight the need for all functional food manufacturers to employ appropriate process and product monitoring and control systems. Such systems must also incorporate procedures to deal with process malfunction and product not conforming to specification. In the case of product evaluation, this would require as a minimum some form of inspection/control system before the product was dispatched. With regard to functional foods, this should involve some form of positive release system based on product analysis.

As part of their quality assurance and control procedures, manufacturers of functional foods will therefore have to be able to demonstrate that the functional ingredient is not only present in the product at the time of manufacture but throughout its shelf-life (ie up to the point at which it is consumed by the end-user). It is implicit in this requirement that the ingredient should not only be present, but be present in sufficient quantities to be able to exert the desired biological effect. This means that, in common with the rest of the food industry, and as a minimum requirement, functional food manufacturers will need to have an ongoing shelf-life evaluation programme. This may involve testing every production run and ensuring that under representative test conditions all attributes of the food were maintained during the shelf-life of the product. Such a programme would require a regular analysis of the data to identify any trends which might affect product quality during this period. In the case of foods containing more fragile functional ingredients, eg probiotic-containing foods, the programme would have to be extended to ensure that the product was meeting specification at the point of delivery to the end-consumer. This may even necessitate some sort of store surveillance exercise as well.

   2.2.3. Product Safety. An essential quality attribute of any food is that it is safe. Food-manufacturing quality management schemes must have systems in place, which as a minimum: identify hazards inherent in the manufacturing system

•   assess the risk of those hazards occurring
•   institute procedures to minimize the risk of hazards occurring
•   monitor (and where necessary improve) the efficacy of those procedures.

   One system which achieves these objectives is Hazard Analysis Critical Control Point (HACCP). The HACCP system was developed by the Pillsbury Company in conjunction with the National Aeronautics and Space Administration (NASA) to ensure food safety for the US manned space programme. For obvious reasons, it was undesirable for astronauts to develop food poisoning on extended space missions! A full discussion of HACCP goes beyond the scope of this chapter. However in summary, the system follows seven basic principles:[17,18]

(i) development of a full understanding of the manufacturing process, this includes an identification of the hazards associated with each step of the process, an assessment of the risk of those hazards occurring and determination of control options

(ii) identification of critical control points (a critical control point might be defined as a point in the manufacturing process where, if appropriate control was not applied, the customer would be put at an unacceptable risk towards a hazard)

(iii) setting of critical limits to ensure that each critical control point is under control

(iv) creation of a critical control point monitoring system by scheduled testing or observation

(v) installation of corrective action procedures in the event that monitoring indicates that a critical control point is moving out of control

(vi) establishment of documentation concerning all procedures and records appropriate to these principles and their application

(vii) verification, on a regular basis, that HACCP is effectively working.

   Experience of food manufacturing processes indicates the importance of ensuring the integrity of all raw materials, together with that of the intermediate and final products in any hazard or risk analysis. For functional ingredients this will necessitate, amongst other things, ensuring that microbial starter cultures have, and keep, the desired genetic characteristics and are free from contamination. The problem of contamination not only applies to microbial cultures. Contamination can also be physical or chemical in nature and applies to all ingredients for example, plant extracts and also seemingly pure chemicals.

An example of the consequences of failure in the latter was the outbreak of eosinophilia-myalgia syndrome associated with the consumption of some tryptophan and tryptophan containing supplements.[19] This appeared to be due to a contaminant in one particular brand of tryptophan.[20] A further point in this area which needs to be considered in the safe manufacture of any functional food is the pharmacological property of the 'functional' ingredient. The market for functional foods is diverse and can include foods to which either pure (single chemical) or impure (eg plant extracts) functional ingredients are added. The dictum of Paracelsus, roughly translated as, *'all things are poisons, it is merely a question of dose'* should not be forgotten. In some cases, the dividing line between an ingredient acting beneficially and detrimentally can be very small. An example of this is hypervitaminosis D in infants associated with the consumption of some vitamin D supplemented foods.[21]

Functional foods can therefore contain ingredients which are either contraindicated for certain groups of the population, or can be hazardous if taken in excessive doses. Thus manufacturers of functional foods will have to ensure that consumers have sufficient details to inform themselves as to the risks of any potential hazards (eg through labelling) and, if necessary, appropriate packaging (eg child-proof tops).

3 CONCLUSION

The increasing awareness of the role of diet in health and the financial premiums associated with foods targeted at the health needs of the population as a whole, or subsections of it, suggest that functional foods will play a greater part in the Nation's diet. In order to avoid charges of 'quackery' and maintain credibility, manufacturers of functional foods will, in common with other food manufacturers, have to demonstrate that their product is safe and consistent. In addition they will also have to demonstrate that their product actually functions. Furthermore, manufacturers of functional foods must be aware of those sub-populations for whom consumption of their product may be contraindicated or may be put at risk by over consumption. This awareness must be translated into appropriate labelling and packaging. These requirements will only be met by employing the appropriate quality assurance systems.

4 ACKNOWLEDGEMENTS

I thank my collegues, Ms. Gráinne Power, Mr. Stanley Cauvain, Dr. Steven Walker and Mr. David Nicholson, for reading the manuscript and their comments.

5 REFERENCES

1.   Secretary of State for Health, 'The Health of The Nation: A Consultative

Document for Health in England,' HMSO, London 1991.

2.   Secretary of State for Health, 'The Health of The Nation: A Strategy for Health in England,' HMSO, London 1992.

3.   D. Maclean, 'Fifty Years of the National Food Survey (Ed. J.M. Slater),' HMSO, London, Chapter 1, p. 3.

4.   T. Ichikawa, 'Functional Foods (Ed. I. Goldberg),' Chapman & Hall, New York, Chapter 18, p. 453, 1994.

5.   Ministry of Agriculture Fisheries and Food, Draft Guidelines on Health Claims on Foodstuffs, Letter to Interested Parties, 1996.

6.   SI 1996, No 1499, The Food Labelling Regulations 1996

7.   British Standards Institution, 'BS 4778 Quality Vocabulary, Part 1 : 1987 International Terms,' 1987.

8.   British Standards Institution, 'Quality Systems, BS EN ISO 9001 : 1994,' 1994

9.   European Food Safety Inspection Service, 'EFSIS Accreditation: Service Protocol, Audit Scope and Guidance Notes (Issue 2),' 1994, Campden & Chorleywood Food Research Association/Meat & Livestock Commission, Milton Keynes.

10.   Food and Agriculture Organization, *'Requirements of Vitamin A, Iron, Folate and Vitamin $B_{(12)}$,* Report of a Joint FAO/WHO Expert Consultation, Food and Nutrition Series No. 23,' FAO, Rome, 1988.

11.   Department of Health, 'Report on Health and Social Subjects No. 46, Nutritional Aspects of Cardiovascular Disease,' HMSO, London, 1994.

12.   C.M. Ripsin, J.M. Keenan, D.R. Jacobs *et al.*, *JAMA.*, 1992, **267**, 3317.

13.   J.M.T. Hamilton-Miller, *Nutrition Bulletin*, 1996, **21**, 199

14.   R. Fuller, *Nutrition Bulletin*, 1996, **21**, 204.

15.   S.E. Gilliland and M.L. Speck, *J. Food Protection*, 1977, **40**, 760.

16.   J.M.T. Hamilton-Miller, *BMJ*, 1996, **312**, 55.

17.   Campden and Chorleywood Food Research Association, 'HACCP: A Practical Guide. Technical Manual No.38' Campden & Chorleywood Food Research Association, Chipping Campden, 1992.

18.   Codex Alimentarius Commission, 'Codex Guidelines for The Application of The Hazard Analysis Critical Control Point (HACCP) System. Joint FAO/WHO Codex Committee on Food Hygiene' WHO/FNU/FOS/93.3 Annex II, 1993

19.   P.A., Hertzman, W.L. Blevins, J. Mayer *et al.*, *New Eng. J. Med.*, 1990, **322**, 869.

20.   E.A. Belongia, C.W. Hedberg, G.J. Gleich, *et al.*, *New Eng. J. Med.*, 1990, **323**, 357.

21.   Department of Health and Social Security, 'Report on Health and Social Subjects No. 15, Rickets and Osteomalacia,' HMSO, London, 1980.

VIABILITY OF *BIFIDOBACTERIA* IN FERMENTED MILK PRODUCTS

J.F. Payne, A.E.J. Morris and P.J. Beers

Food Research Centre
University of Lincoln Grimsby Campus
61 Bargate
Grimsby DN34 5AA

1 INTRODUCTION

The five most commonly occurring bifidobacteria in the human digestive tract are *Bifidobacterium bifidum, B. longum, B. adolescentis, B. infantis* and *B. breve*. Of these, *B. longum* and *B. bifidum* are most frequently selected for use in foods and in particular in yoghurts. Bifidobacteria are often described as being probiotic as they are thought to have beneficial effects on the health of the host. Diet, antibiotic therapy, stress and other factors may disrupt the delicate balance of the gastrointestinal microflora. Consumption of yoghurt containing bifidobacteria helps to restore this balance in favour of beneficial microorganisms. Bifidobacteria produce both lactic and acetic acid in a ratio of 2:3, which lowers the pH in the large intestine, making the conditions unfavourable for the growth of many putrefactive and pathogenic organisms. Bifidobacteria have also been reported to compete for colonisation space in the intestine[1] and preparations have been used in the treatment of diarrhoea in children and constipation in the elderly. *In vitro* and animal tests have demonstrated the infection prevention, immunity activation, anti-tumorigenic effects and vitamin production of bifidobacteria.[2]

In order to have therapeutic effects, yoghurts should contain bifidobacteria of human origin, and should be able to survive digestion, and capable of surviving and growing in the intestine, producing beneficial effects for the host.[3] As bifidobacteria are noted for their low acid and oxygen tolerance[4] and nutritional fastidiousness, the challenge for manufacturers is to develop products in which a therapeutic dosage is maintained to the end of the shelf life. This is considered to be a minimum of 10^6 viable bifidobacteria per gram, and 150 grams should be consumed on a regular basis.[5] An additional complication is the fact that several media have been reported as suitable for the selective enumeration of bifidobacteria in yoghurts, but as yet there is no consensus of opinion as to which will give the most accurate results for a range of products. Selection of the wrong media or technique (spread or pour plate) may result in low viable counts being achieved.

The aims of this work were therefore:

      (i)   to determine whether a therapeutic dose of three species of bifidobacteria could be maintained in yoghurt for a 21-day storage period

(ii)   to compare viable counts of bifidobacteria obtained utilising pour and spread plate techniques and four different selective media for isolation of *Bifidobacterium* sp.

(iii)   to recommend the most appropriate medium and technique, for use by manufacturers in assessment of viable counts of bifidobacteria, in mixed cultures in yoghurt.

2 MATERIALS AND METHODS

Freeze dried pure cultures of *Bifidobacterium bifidum* NCTC 10472, *B. longum* NCTC 11818 and *B. adolescentis* NCTC 11814 were obtained from the National Collection of Type Cultures (NCTC), Central Public Health Laboratory, London and a frozen, DVS yoghurt culture (Type B3) containing *Lactobacillus delbreuckii* ssp. *bulgaricus* and *Streptococcus salivarius* ssp. *thermophilus* was obtained from Chr. Hansen's Laboratory, Reading. After resuscitation bifidobacteria cultures were maintained in NGYC broth.[6]

Spray dried semi-instantised skimmed milk powder was reconstituted in distilled water to 12% (w/v), heat treated to 85°C for 30 minutes and then cooled to 37°C. In order to prepare inocula for experimental purposes, 70ml aliquots of this medium were inoculated with NGYC bifidobacteria culture at a rate of 10% (v/v) and incubated anaerobically at 37°C until the milk set. The inocula were then stored at 4°C prior to use.

Reinforced Clostridial Agar (RCA) (Oxoid, UK) was used to enumerate bifidobacteria in the inoculum, using both spread and pour plate techniques. Plates were incubated anaerobically at 37°C for 72 hours. The yoghurt culture inoculum was enumerated by the pour plate technique, using M17 agar (Oxoid, UK) for *S. salivarius* ssp. *thermophilus* and MRS agar (Oxoid, UK) acidified to pH 5.4 for *L. delbreuckii* ssp. *bulgaricus*. The M17 plates were incubated aerobically and the MRS anaerobically, both at 37°C for 72 hours. Table 1 shows the viable counts obtained for each of the starter culture organisms and the concentration of viable cells in the milk on inoculation with starter.

Two 1.5 litre aliquots of reconstituted skimmed milk in 2-litre flasks were prepared and heat treated as described above. They were then inoculated with 1.5% (v/v) *Bifidobacterium* culture, and 0.5% (w/v) B3 yoghurt culture. After thorough mixing the flasks were incubated at 37°C. The pH was measured at hourly intervals, until it had dropped to 4.6 (approximately four to six hours), after which the yoghurt was decanted into 150ml pots with heat-sealed, foil lids and stored at 5°C.

Duplicate samples from each batch were analysed for pH, total acidity (Table 2) and viable count of bifidobacteria using four selective media. The analyses were carried out immediately after production and after seven, 14 and 21 days storage. Enumeration involved making a dilution series in 0.1% peptone water, followed by duplicate plating by both spread and pour plate techniques onto Modified Rogosa's agar (RMS),[7] NPNL agar,[8] AMC agar[9] and MRS+NNPL agar.[10] The plates were incubated anaerobically at 37°C for 72 hours. The results given in Tables 3, 4 and 5 are thus the means of a total of four separate counts.

Table 1 *Starter Culture Viable Count and Concentration in Milk on Inoculation*

Starter culture organism	Inoculum viable count (colony forming units/ml)	Concentration in milk on inoculation (colony forming units /ml)
B. bifidum	6.6×10^{10}	9.9×10^8
S. salivarius ssp. *thermophilus*	1.37×10^{11}	6.85×10^8
L. delbreuckii ssp. *bulgaricus*	7.75×10^9	3.88×10^7
B. longum	9.3×10^9	1.4×10^8
S. salivarius ssp. *thermophilus*	1.37×10^{11}	6.85×10^8
L. delbreuckii ssp. *bulgaricus*	7.75×10^9	3.88×10^7
B. adolescentis	1.78×10^8	2.67×10^6
S. salivarius ssp. *thermophilus*	4.4×10^9	2.2×10^7
L. delbreuckii ssp. *bulgaricus*	3.6×10^9	1.8×10^7

3 RESULTS AND DISCUSSION

The microbiological media used in this study were selected on the basis of results obtained by the authors, (unpublished), using pure cultures, in which 10 microbiological media were assessed for both recovery of bifidobacteria and inhibition of standard yoghurt culture organisms and *Lactobacillus acidophilus*. Of those microbiological media evaluated AMC, RMS and NPNL were found to be the most successful and were therefore selected for further work in this investigation.

AMC agar was adapted from LP agar[11] and mBIM-1[12] and was recommended for the selective enumeration of *B. longum* from mixed lactic culture. Samona and Robinson (1991),[13] compared a number of media for their ability to selectively enumerate bifidobacteria in the presence of other lactic cultures, and of these recommended RMS to monitor survival of bifidobacteria in a range of dairy products. NPNL originally developed by Teraguchi *et al.* (1978)[14] and translated into English by Laroia and Martin (1991),[8] has been used successfully by a large number of workers for the enumeration of bifidobacteria in both pure and mixed culture studies. MRS + NNLP [10] was also included in this study as this is the medium recommended by Christian Hansen's for the selective enumeration of bifidobacteria in yoghurt.

Yoghurts prepared using mixed cultures as described above were stored at 5°C. As expected their pH dropped with a concomitant rise in total acidity, as a result of post-incubation acidification (Table 2). The greatest decrease in pH occurred during the first seven days of storage, and for most of the storage period it was well below 4.3. Shah (1997)[4] states that due to the low acid tolerance of bifidobacteria their numbers decline rapidly in yoghurt, and Ishibashi and Shimamura (1993)[2] reported the adverse effects of post-incubation acidification on cell viability. In contrast, Samona and Robinson (1994)[15] found that although yoghurt cultures adversely affected the growth of *Bifidobacterium sp.* in milk, they had little effect on their long term viability, and that a pH above 4.0 had no marked effect. This is in agreement with the results shown in

Table 2 *Changes in pH and Total Acidity (T.A.) in Yoghurts During 21 days Storage at 5°C*

Bifidobacterium sp. in starter culture	pH / Total Acidity (% Lactic acid)	Day 0	Day 7	Day 14	Day 21
B. bifidum	pH	4.58	4.3	4.24	4.25
	T.A.	1.17	1.37	1.45	1.45
B. longum	pH	4.53	4.18	4.20	4.24
	T.A.	1.16	1.51	1.53	1.52
B. adolescentis	pH	4.58	4.28	4.24	4.24
	T.A.	1.16	1.36	1.43	1.46

Tables 3, 4 and 5, for AMC and RMS agars, which demonstrate that although there was a decline in numbers during the storage period, a minimum therapeutic dosage of at least 10^6 cfu/g for each of the *Bifidobacterium* sp. was maintained. However, when assessed using both NPNL and MRS+NNLP agars all the yoghurts contained bifidobacteria at levels below the required minimum therapeutic dose.

The results obtained using AMC medium indicated that the number of organisms detected also depended on the plating technique employed. The pour plate technique gave consistently higher counts than the spread plate technique for each of the *Bifidobacterium* sp. examined, and the difference in recovery between the pour and spread plate techniques was consistent throughout the trial (Tables 3-5). All of the cultures were incubated in an anaerobic environment making it unlikely that it was due to oxygen tension.

This is an area that requires more investigation. Conversely, the results obtained using RMS medium were found to be comparable for both plating techniques with a minimum therapeutic dose indicated for all *Bifidobacterium* sp. throughout the trial, with the exception of the Day 21 pour plate count for *B. adolescentis* of 9.43×10^5.

There is a considerable degree of variation in the constituents of the four microbiological media used in this study. Reducing agents are added to microbiological media for the culture of anaerobic organisms in order to reduce the redox potential and thus, promote growth. Bifidobacteria are unable to grow in a medium with a high redox potential and hence, reducing substances are added to assist in obtaining the anaerobic conditions required.[16] Examples of such substances contained within the microbiological media are: glucose, present in AMC, NPNL and MRS+NNLP agars, which when heated is said to increase in reducing capacity[17] and L-cysteine-HCl, present within the formulations of AMC, RMS and NPNL agars. This substance not only reduces the redox potential but is also recognised as the only essential amino acid required for growth by bifidobacteria.[17,18]

The microbiological media also contain bifidogenic factors ie substances which promote the growth of bifidobacteria. One such bifidogenic factor is lactulose, an isomer of lactose formed when milk is heated.[17b] RMS is the only medium of those used to contain lactose, (10g/L). It would be expected that a proportion of this would be converted to lactulose when the medium was autoclaved (121°C for 15 minutes). Other growth factors such as bifidus factor 2, peptides derived from the enzymatic hydrolysis

Table 3 *Recovery of B. bifidum From Yoghurt Over a 21-day Storage Period*

Medium	Plating Technique	Day 0 (cfu/g)	Day 7 (cfu/g)	Day 14 (cfu/g)	Day 21 (cfu/g)
AMC	Spread	$< 10^5$	$< 10^5$	$<10^4$	$<10^4$
	Pour	2.75×10^5	3.05×10^8	6.45×10^8	5.17×10^8
RMS	Spread	1.35×10^8	1.28×10^8	9.17×10^8	2.71×10^7
	Pour	3.08×10^8	2.27×10^8	2.99×10^8	1.31×10^6
NPNL	Spread	$< 10^5$	$< 10^5$	$<10^4$	$<10^4$
	Pour	1.65×10^5	$< 10^4$	$<10^3$	$<10^3$
MRS+NNLP	Spread	$< 10^5$	$< 10^5$	$<10^4$	$<10^4$
	Pour	1.80×10^5	$< 10^4$	3.25×10^4	$<10^3$

Table 4 *Recovery of B. longum From Yoghurt Over a 21-day Storage Period*

Medium	Plating Technique	Day 0 (cfu/g)	Day 7 (cfu/g)	Day 14 (cfu/g)	Day 21 (cfu/g)
AMC	Spread	3.43×10^7	$< 10^5$	$<10^4$	$<10^4$
	Pour	1.98×10^8	2.93×10^8	7.13×10^8	3.33×10^8
RMS	Spread	3.02×10^8	2.36×10^8	7.51×10^7	8.3×10^7
	Pour	1.20×10^9	3.55×10^8	6.15×10^8	4.75×10^8
NPNL	Spread	2.16×10^9	$< 10^5$	3.50×10^4	$<10^4$
	Pour	5.46×10^7	$< 10^4$	$<10^3$	$<10^3$
MRS+NNLP	Spread	$< 10^5$	$< 10^5$	$<10^4$	$<10^4$
	Pour	$< 10^4$	$< 10^4$	$<10^3$	$<10^3$

Table 5 *Recovery of B. adolescentis From Yoghurt Over a 21-day Storage Period*

Medium	Plating Technique	Day 0 (cfu/g)	Day 7 (cfu/g)	Day 14 (cfu/g)	Day 21 (cfu/g)
AMC	Spread	$<10^5$	$<10^5$	$<10^4$	$<10^4$
	Pour	1.0×10^7	5.30×10^8	4.38×10^8	3.77×10^8
RMS	Spread	2.21×10^7	3.69×10^7	9.13×10^6	9.53×10^6
	Pour	1.05×10^8	3.93×10^8	2.53×10^8	9.43×10^5
NPNL	Spread	9.5×10^5	$<10^5$	$<10^4$	$<10^4$
	Pour	1.15×10^6	$<10^4$	$<10^3$	$<10^3$
MRS+NNLP	Spread	$< 10^5$	$<10^5$	$<10^4$	$<10^4$
	Pour	$< 10^4$	$<10^4$	$<10^3$	$<10^3$

of proteins,[17b] are also present in all the media in the form of yeast extract or autolysate and additionally, in the form of Tryptone in both RMS and NPNL agars.

4 CONCLUSIONS

Provided the concentration of bifidobacteria cells/ml inoculated milk is of the order of 10^8 to 10^9 it should be possible to maintain the therapeutic dosage throughout a 21-day refrigerated storage period. These results demonstrate that this was possible without the use of special barrier pots to maintain anoxic conditions.

Of the media examined RMS and AMC were by far the most effective in recovering bifidobacteria. If AMC is to be used it is important that the pour plate technique is utilised as opposed to the spread plate technique which yielded much lower viable counts.

RMS was found to be equally effective for the isolation of *B. bifidum*, *B. longum* and *B. adolescentis* which are the species most commonly used in dairy products. Thus, the most appropriate selective medium for enumeration of bifidobacteria in yoghurt was found to be RMS using either the spread plate or pour plate technique.

5 REFERENCES

1. D. G. Hoover, *Food Technol.*, 1993, **47** (6),120.
2. N. Ishibashi and S. Shimamura, *Food Technol.*, 1993, **47** (6), 129.
3. S. E. Gilliland, *J. Dairy Sci.*, 1989, **72,** 2483.
4. N. P. Shah, *Milchwissenschaft*, 1997, **52** (1), 16.
5. R. K. Robinson, *S. Afri. J. Dairy Sci.*, 1990, **22**, (2), 43.
6. W. E. V. Lankaputhra, N. P. Shah, and M. L. Britz, *Milchwissenschaft*, 1996, **51,** (2), 65.
7. K. Shimada, M. Mada, M. Mutai, A. Suzuki and H. Konuma, *J. of Food Hygienic Soc. of Japan*, 1977, **18,** 537.
8. S. Laroia and J. H. Martin, *Cult. Dairy Prod. J.*, 1991, **26** (2), 32.
9. L. Arroyo, L. N. Cotton and J. H Martin, *Cult. Dairy Prod. J.*, 1995, **30** (1), 12.
10. Anon., *Nutrish Cultures,* Chr. Hansen.
11. L. Lapierre, P. Undeland and L. J. Cox, *J. Dairy Sci.*, 1992, **75**, 1192.
12. F. J. Munoa and R. Pares, *Appl. Environ. Microbiol.*, 1988, **54** (7), 1715.
13. A. Samona and R. K. Robinson, *J. of the Soc. of Dairy Tech.*, 1991, **44** (3), 64.
14. S. Teraguchi, M. Uehara, K. Ogasa and K. Mitsuoka, *Jpn. J. Bacteriol.*, 1978, **33,** 753.
15. A. Samona and R. K. Robinson, *J. of the Soc. of Dairy Tech.*, 1994, **47** (2), 58.
16. D. Roy, F. Dussault and P. Ward, *Milchwissenschaft*, 1990, **45,** (8), 500.
17. J. Rasic and J. Kurmann, '*Bifidobacteria and Their Role.*' Birkhauser Verlag Basel, Switzerland, 1983, Chapter 9, p.144.
17b J. Rasic and J. Kurmann, '*Bifidobacteria and Their Role.*' Birkhauser Verlag Basel, Switzerland, 1983, Chapter 4, p.42.
18. J. B. Hassinen, G. T. Durbin, R. M. Tomarelli and F. W. Bernhart, *J. of Bacteriol.*, 1951, **62,** 771.

MICROENCAPSULATION OF MARINE OILS WITH A VIEW TO FOOD FORTIFICATION

M.E. Ní Néill and K.M.Younger

Department of Biological Sciences
Dublin Institute of Technology
Dublin 8, Ireland

1 INTRODUCTION

Population and clinical studies have demonstrated health benefits and reduced mortality from cardiovascular disease among people consuming fish.[1,2] Benefits have been attributed to the omega-3 (n-3) fatty acids in fish. As well as being low in total and saturated fat, fish is known to be a rich source of the unique long chain polyunsaturated fatty acids of the omega-3 family, including both eicosapentaenoic acid (EPA, 20:5n-3) and docosahexaenoic acid (DHA, 22:6n-3). Large doses of omega-3 fatty acids may have ameliorative effects in hypertension, inflammatory, immune and other diseases. Research from epidemiological studies in different populations,[3,4] clinical trials in patients,[5] and healthy subjects,[6] animal experiments,[7] biochemical studies and cell culture experiments[8] have expanded our understanding of how these substances contribute to health. Several health benefits have been associated with the regular consumption of omega-3 fatty acids. Two of these—heart health and prenatal development—are especially germane to consumer interests because of the large number of people potentially affected.

Many people find fatty fish and marine oils unpalatable and are reluctant to include them in their regular diet. Polyunsaturated marine oils, particularly n-3 fatty acids, are also susceptible to oxidative deterioration, limiting their usage in foods.

Microencapsulation is a process in which small amounts of liquids, solids or gases (core) are coated with materials (coating) which provide a barrier to undesirable environmental and/or chemical interactions (eg heat, moisture, oxidation) until release is desired.[9] Microencapsulation masks tastes, flavours and odours, provides protection against oxidation and puts active ingredients into a free flowing powder for ease of handling and incorporation into dry food systems.[10]

Spray drying is the most common method of encapsulating food ingredients.[11] In the preparation of microencapsulated spray dried powders an emulsion is made by dissolving the microencapsulating coating materials in water, adding the core and making a fine emulsion of <2μm in particle size. The emulsion is then spray dried. Important requirements in the selection and evaluation of coating materials are:[12-13]

- high solubility

- low viscosity at high solids content
- good emulsion stability and film forming properties
- good oil retention in the spray dried microcapsules
- low extractable surface oil
- protection from oxidation.

Our objective was to develop a methodology for the microencapsulation of marine oils using Irish materials where possible. Microcapsules were analysed to determine the suitability of various formulations for the fortification of foods not normally associated with fish with sufficient quantities of marine oils to confer a health benefit (350mg n-3 fatty acids/day).

2   MATERIALS AND METHODS

2.1 Materials

2.1.1 Wall and Core Materials. Skimmed milk powder (SMP) and sodium caseinate (NaC) were obtained from Virginia Milk Products, Co. Cavan. Whey powder (WP) and whey protein concentrate (WPC) were obtained from Dairygold Foods, Co. Cork. Modified starch (N-Lok) was obtained form National Starch and Scientific, Manchester. Maltodextrin of DE 19 (Glucidex 19D) was obtained from Corcoran Chemicals Ltd, Dublin. Samples of cod liver oil were supplied by Rice Steele.

2.1.2 GC Standards. Fatty acid methyl esters (10:0,12:0,14:0,16:0,16:1,18:0,18:1, 18:2,18:3,18:4,20:1,20:4,20:5,22:1,22:4,22:6) were purchased from Sigma-Aldrich Ireland Ltd, Dublin.

2.2 Microencapsulation

2.2.1 Emulsion Preparation. Aqueous solutions of wall materials were prepared (composition, Table 1) with blending and heating if required, and stored overnight at 4°C to ensure complete disolution. Cod liver oil was emulsified into the wall solutions at a level of 4:1 (wall:core). Emulsification was carried out by first preparing a coarse emulsion using a Silverson medium viscosity liquid mixer (Armfield Technical Education Co., Hampshire, England), equipped with a fine emulsor screen and operated at half maximum speed for one minute, followed by three successive homogenisation steps using an Armfield Laboratory Pressure Homogeniser (Armfield Technical Education Co., Hampshire, England) operated at 1,000 psi. The final emulsion was spray dried within 30 minutes.

2.2.2 Spray Drying. Emulsions were spray dried using an APV Anhydro Laboratory Spray Dryer (APV Anhydro A/S, Søborg, Denmark). The dryer had an evaporation rate of 7.5kg/hour and a chamber diameter of 1 metre. The emulsions were fed into the spray dryer by a peristaltic pump at a flow of 30g/minute. The centrifugal atomiser was operated at 180 Volts and drying was in the concurrent configuration. Inlet air at 155°C and oulet air at 80-85°C were used in all cases.

Table 1 *Formulations for Preparing Marine Oil Microcapsules*

Code	Components	Solids content	Wall:Core
50% SMP	Skimmed milk powder	50%	4:1
35% SMP	Skimmed milk powder	35%	4:1
50% Whey P	Whey powder	50%	4:1
50% WPC	Whey protein concentrate	50%	4:1
50% N-Lok	Modified starch (N-Lok)	50%	4:1
50% MD+NaC (85:15)	Maltodextrin, sodium caseinate	50%	4:1
50% SMP+NaC (85:15)	Skimmed milk powder, Sodium caseinate	50%	4:1
50% MD+WPC (85:15)	Maltodextrin, whey protein concentrate	50%	4:1
32% MD+WPC (9:1)	Maltodextrin, whey protein concentrate	32.5%	3.3:1
50% MD+WPC (9:1)	Maltodextrin, whey protein concentrate	50%	4:1
50% MD+WPC (1:1)	Maltodextrin, whey protein concentrate	50%	4:1

2.3 Physical Analyses

2.3.1 Emulsion Stability. Dilutions (0.1%) of each emulsion were prepared using distilled water, and centrifuged at 500xG. They were evaluated at 0, 5, 10, 20 and 30 munute intervals using a spectrophotometer at 400nm. Absorbance readings were measured immediately following centrifugation. The rate of change in optical density during the first 10 minutes for each emulsion was compared.

2.4 Chemical Analyses

2.4.1 Oil Content. Oil content of the dry microcapsules was determined by a modified Weibull-Berntrop method. 3-4g powder (x2) was accurately weighed into a round bottomed flask, containing a few anti-bumping granules. Water at 30°C was added to give a total volume of 25ml. 20% hydrochloric acid (50ml) was added to this diluted test portion and gently mixed. The mixture was refluxed for 30 minutes. The condenser and the neck of the flask were rinsed into the flask with water (about 150ml) heated to at least 80°C). Acid washed Celite 1g was added, the contents of the flask were immediately filtered and the flask was washed well through this filter using hot water until the washing were acid-free as indicated by blue litmus paper. The filter paper was removed using tongs, placed in a Soxhlet extraction thimble, dried thoroughly in an oven ($103^0C \pm 5^0C$, 8-9 hours), cooled and transferred to a Soxhlet extractor. The fat was extracted from the filter paper using petroleum ether for four hours (approximately five minutes per cycle). A blank determination was simulatenously carried out using the same procedures and reagents but replacing the diluted test portion with 25ml water.[14]

2.4.2 Petroleum Ether Extractable Surface Oil. 4-5g powder (x2) was accurately weighed and pentane (15ml) was added. The solvent-powder mixture was gently shaken at 180min[-1] for 20 minutes. The mixture was filtered into a preweighed dry flask (Whatman no.1 filter paper), the solvent was distilled off and the surface oil on the surface w.r.t. total oil retained in the powder.[15] The ratio (as percentage) of encapsulated ester to initial ester in the emulsion, dry solid basis, was defined as the retention. The percentage of encapsulated ester that could not be extracted from the dry microcapsules was designated the encapsulation efficiency (EE).[13]

2.5 Oxidative Stability

Original oil, emulsion and microcapsule samples were taken during processing. Individual 3g portions of microcapsules were stored 60°C, 100°C, 150°C and 235°C for 30 minutes in open amber glass bottles and samples were taken immediately after heating. All samples were stored in sealed vials under nitrogen at -20°C until analysis.

2.5.1 Sample Extraction for GC-MS. 0.2-2g sample was weighed into a homogenising vessel. Sufficient deionised water was added to bring the total water present to 3.2ml, together with 8ml methanol and 4ml dichloromethane. Sample was homogeneised for two minutes, 4ml dichloromethane was added and homogenised for a further 20 seconds. Deionised water (4ml) was added and homogeneised for 20 seconds. Mixture was centrifuged for 10 minutes at 2,000-2,500 rpm. As much of the lower dichloromethane layer as possible was drawn off, filtered and the solvent was evaporated under a stream of nitrogen.

2.5.2 Transmethylation. Original and extracted oils were transmethylated using a modified AOCS-AOAC method.[16] Approximately 0.1g was weighed into a glass tube (16x125mm, with leak-tight, teflon-lined screw caps.) 0.5N methanolic NaOH (1.5ml) was added, blanketted with nitrogen, capped tightly, mixed, and heated at 100°C for five minutes. The mixture was cooled, BF_3 (2ml, 12% in methanol) was added, blanketted with nitrogen, capped tightly, mixed and heated at 100°C for 30 minutes. The mixture was cooled to 30-40°C, hexane (1ml) was added, blanketted with nitrogen, capped tightly and shaken vigorously for 30 seconds while still warm. NaCl solution (5ml) was immediately added, blanketted with nitrogen, capped tightly and thoroughly agitated. The mixture was cooled to room temperature and allowed to separate into layers. The hexane layer was removed, transferred into a clean autosampler vial (2ml), blanketted with nitrogen and capped tightly. 1-2μl was injected into the GC-MS.

2.5.3 GC-MS. HP 6890 GC-MS system fitted with a Quadrex 007-FFAP-25 capillary column. Operating conditions: Injection port temperature 250°C; MS transfer line 280°C; oven programmes from 120°C (no initial hold), 15°C/min to 200°C, hold 10 minutes, 5°C/minutes to 210°C, final hold 15 minutes. Capillary column injection system: split mode at split ratio of 1:50. Helium carrier flow rate: 0.4ml/minutes.

Fatty acid methyl esters were identified by comparison with standards and using the GC-MS library (NBS75K). EPA and DHA content of the soil samples was measured as a percentage of the stable palmitic acid (16:0) content in each sample (referred to as the palmitic acid ratio). A plot of temperature versus percentage of palmitic acid for EPA and DHA (Figures 1 and 2) indicated the extent of oxidation protection afforded by the microencapsulation process.

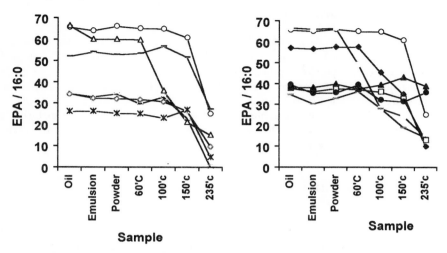

Formulations: Original oil-○- ;50% N-Lok ━━━;50% SMP- -;35% SMP-△-;50% Whey P-◈-;
50% WPC 30-✳— ;50% MD+NaC 85:15-◆- ; 35% SMP+NaC 85:15━ ;
32.5% MD+WPC 9:1-☐-;50% MD+WPC 9:1-▲-;50% MD+WPC 85:15━;
50% MD+WPC 1:1-●- .

Figure 1 *Effect of processing and heating on EPA*

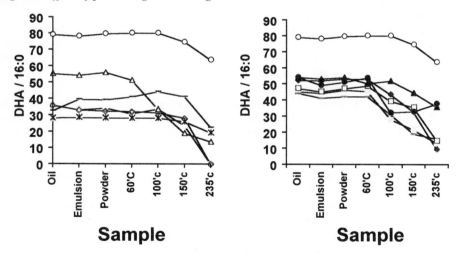

Formulations: Original oil-○- ;50% N-Lok ━━━;50% SMP- -;35% SMP-△-;50% Whey P-◈-;
50% WPC 30-✳— ;50% MD+NaC 85:15-◆- ; 35% SMP+NaC 85:15━ ;
32.5% MD+WPC 9:1-☐-;50% MD+WPC 9:1-▲-;50% MD+WPC 85:15━;
50% MD+WPC 1:1-●- .

Figure 2 *Effect of processing and heating on DHA*

2.6 Fortification and Sensory Analysis

2.6.1 Bread Making. Soda brown bread was fortified with that amount of microcapsules which contained 16.5g oil/900g bread. Ingredients: 52% whole-grain flour, 3% baking powder, 0.5% castor sugar, 1% salt, 43% milk. The microcapsules were sieved into the premixed dry ingredients, the milk was added and the bread mix was baked at 235°C for 30 minutes.

2.6.2 Taste Testing. A triangle taste test was carried out by an untrained panel to determine whether or not the fortification with microencapsulated fish oil had perceptibly altered the bread flavour.

3 RESULTS AND DISCUSSION

N-Lok is a modified starch and dextrin product developed to replace the traditionally used microencapsulation material, gum arabic. It is reported to have excellent emulsion stability, oxidative resistance and encapsulation ability.[17] The performances of the test formulations were compared to that of N-Lok to judge their suitability as wall material systems for the microencapsulation of marine oils. Skimmed milk powder and whey powder are readily available in Ireland and were chosen for testing due to their emulsifying properties. Whey proteins have many of the proteins required from a coating material and have been reported as an effective basis for microencapsulation by spray drying of volatiles.[13] In light of this, whey protein concentrate was also tested. Maltodextrins are non-sweet nutritive polysaccharides produced by the partial hydrolysis of starch. They have virtually no emulsion-stabilising effect other than the viscosity imparted to the emulsion at high solid levels, however high dextrose-equivalence maltodextrins can protect against oxidative deterioration and, when used in a blend with protein, have been reported effective in the microencapsulation of certain oils.[11,18] Initially formulations were tested which contained one coating material only, in order to highlight the strengths and weaknesses of those materials. Formulations were then modified by the addition of other components in order to improve their microencapsulating ability.

3.1 Emulsion Stability

Emulsion stabilities of various wall material formulations were compared (Figure 3). Of the single component formulations, 50% WPC produced the most stable emulsions. In their natural states, whey protein concentrates have a higher emulsification capacity than skimmed milk powder or casein.[19] The 50% SMP and 50% Whey Powder emulsions were less stable than the 50% N-Lok emulsion. All other formulations produced more stable emulsions, the most stable being those with sodium caseinate or whey protein concentrate.

3.2 Oil Retention

The retention of oil by the various wall systems after microencapsulation by spray drying was compared (Table 2). 50% N-Lok microcapsules retained only 55.08% of the oil added to the inital emulsion. Single component formulations containing skimmed milk powder gave very poor retentions, 50% whey powder and 50% WPC retained a higher proportion of added oil than the N-Lok, 61.0% and 74.2% respectively. Addition of sodium caseinate to skimmed milk powder at 15% solids level had an adverse effect on the oil retention, resulting in only 17.3% of the added oil being retained. Comparison of oil retention, resulting in only 17.3% of the added oil being retained. Comparison of different 50% MD+EPC formulations showed an increase in retention with increasing

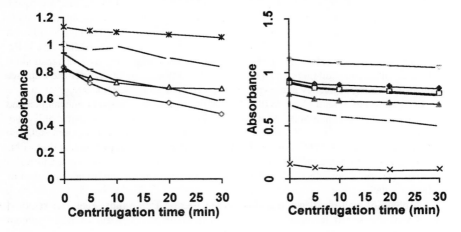

Formulations: 50% N-Lok ▬; 50% SMP – –; 35% SMP △; 50% Whey P ◆;
50% WPC 30 ✱;50% MD+NaC 85:15 ◆; 35% SMP+NaC 85:15 —;
32.5% MD+WPC 9:1 ☐;50% MD+WPC 9:1 ▲;50% MD+WPC 85:15 ▬;
50% MD+WPC 1:1 ● ;35% MD+SMP 85:15 —.

Figure 3 *Emulsion stability of formulations*

Table 2 *Oil Retention, Uncapsulated Surface Oil and Encapsulation Efficiency of Various Wall Systems*

Formula	Oil retained (g/100g)	Unencapsulated surface oil (g/100g)	Encapsulated oil (g/100g)	Oil Retention (%)	Encapsulation Efficiency (%)
50% N-Lok	11.016	0.181	10.835	55.08	98.357
50% SMP	4.320	4.116	0.204	21.60	4.722
35% SMP	8.551	2.969	5.582	42.755	65.279
50% Whey P	12.200	1.485	10.715	61.00	87.828
50% WPC 30	14.836	0.482	14.354	74.18	96.751
50% MD+NaC 85:15	9.468	0.565	8.903	47.34	94.033
35% SMP+NaC 85:15	3.452	1.706	1.746	17.26	50.565
32.5% MD+WPC 9:1	12.630	2.215	10.415	42.1	82.462
50%MD+WPC 9:1	10.076	0.233	9.843	50.38	97.688
50% MD+WPC 85:15	11.53	0.880	10.65	57.65	92.368
50% MD+WPC1:1	14.32	1.444	12.876	71.6	89.916

WPC content, up to 71.6% retention with the 50% MD+WPC 1:1 formulation. These results demonstrate the effect that the emulsion stabilising ability of the wall system has on the retention of core. Formulations with low emulsion stability may be unable to keep the dryer infeed emulsion well dispersed during the drying operation, allowing the emulsion break in the feed line or surge tank of the dryer.

3.3 Unencapsulated Surface Oil and Encapsulation Efficiency

The proportions of oil that could not be extracted from the dry microcapsules (encapsulation efficiency) were compared after measuring the extractable surface oil of each formulation (Table 2). Ideally, encapsulating efficiency should be >99%. The 50% N-Lok microcapsules had only 0.2% of the added oil exposed in the surface of the microcapsules, resulting in an encapsulation efficiency of 98.4%, only slightly falling short of the ideal. The initial single-component formulations had quite high surface oil contents, ranging from 1.5% for 50% whey powder to 4.1% for 50% SMP. 50% WPC was the exception with only 0.5% unencapsulated surface oil and encapsulation efficiency of 96.8%. Addition of sodium caseinate to skimmed milk powder did improve the encapsulating efficiency somewhat, but not to an acceptable level.

Comparison of different MD+WPC formulations shows an increase in unencapsulated surface oil with increasing WPC content, resulting in a drop in encapsulation efficiency from 97.7% with 10% WPC to 89.9% with 50% WPC. The highest encapsulation efficiency was given by 50% MD+WPC 9:1, which at 97.7% almost equals that of N-Lok. Sheu and Rosenberg[13] also found a 9:1 ratio of high DE maltodextrin and whey protein isolates to give the best encapsulation efficiency in the microencapsulation of ethyl caprylate.

A blend of maltodextrin and sodium caseinate at a 15% solids level only retained 47.3% of the added oil, less than the same ratio of maltodextrin to whey protein concentrate, although the encapsulation efficiencies were similar.

3.4 Oxidative Stability

A series of heating tests was carried out on each of the dry microcapsules produced to compare the oxidative protection afforded by the different coating material formulations. The fatty acid compositon of the microencapsulated oil was analysed before and after the drying process in order to eliminate any changes caused by the processing step. The same procedure was carried out on unencapsulated cod liver oil. The effect of processing and heating on EPA content to the oils is shown in Figure 1. The sharp drop in the palmitic acid ratio shown by the unencapsulated cod liver oil above a temperature of 150°C is also observed for most of the microencapsulated oils, except in the case of 50% MD+WPC 9:1 and 50% MD+WPC 1:1. Both of these formulations display a protective effect at the high temperatures used. The effect of processing and heating DHA is displayed in Figure 2. The same protective effect is not seen for the DHA.

3.5 Sensory Analysis

Triangular taste and preference tests were carried our using an untrained panel of volunteers. In all cases, the triangular taste tests indicated that there was a detectable difference between the fortified and the unfortified breads. To determine if this difference was acceptable, paired preference tests were conducted. Again, all tests showed that the preference was the unfortified bread.

4 CONCLUSIONS

Marine oil microcapsules were successfully prepared in terms of oil retention and encapsulation efficiency. Formulations with highest WPC content had the best oil retention. WPC formulations also had the best encapsulation efficiency.

Microencapsulation afforded some protection to EPA against oxidative deterioration, but not to DHA. Formulations with highest Maltodextrin content provided the best oxidative protection.

Food fortification was unsuccessful due to unacceptable sensory attributes of the fortified bread. It is possible that the soda brown bread was an unsuitable vehicle for microencapsulated marine oils and that future work may find the microcapsules more acceptable in another food.

5 REFERENCES

1. J.A. Nettleton, 'Seafood Science and Technology', Ed. E.G. Bligh, Blackwell Scientific Publications Inc., Massachusetts, 1992, Chapter 4, p.32.
2. J.B. Holub, *Can. Med. Assoc. J.*, 1998, **139**, 377.
3. Y. Kawaga *et al.*, *J. Nut. Sci. Vitaminol.*, 1982, **28**, 441.
4. D. Kromhout, E.B. Bosscheiter and C. de Lezenne Coulander, *N. Engl. J. Med.*, 1985, **312**, 1205.
5. B.E. Phillipson, D.W. Rothrock, W.E. Connor, W.S. Harris and D.R Illingworth, *New Engl. J. Med.*, 1985, **312**, 1210.
6. A.M. Fehily, M.L. Burr, K.M. Phillips and N.M. Deadman, *Am. J. Clin. Nutr.*, 1983, **38**, 349.
7. H.R. Davis, R.T. Bridenstine, D. Vasselinovitch and R.W. Wissler, *Arteriosclerosis*, 1987, 7, 441.
8. T.H. Lee, R.L. Hoover, J.D. Williams, R.I. Sperling *et al.*, *N. Eng. J. Med.*, 1985, 312, 1217.
9. H.C. Greenblat, 'Encapsulation and Controlled Release', Ed. D.R. Karsa and R.A. Stephenson, Royal Soc. of Chem., Cambridge, 1993, p.148.
10. C. Thies, 'Encyclopaedia of Polymer Science and Engineering', Ed. H.F. Mark *et al.*, Interscience Publishers, N.Y., 1987, Vol. 9, p,148.
11. W.E. Bangs and G.A. Reineccius, *J. Food Sci.*, (1981), **47**, 254-259.
12. W.E. Bangs and G.A. Reineccius, *J. Food Sci.*, 1990, **55** (5), 1356-1358.
13. T.Y. Sheu and M. Rosenberg, *J. Food Sci.*, 1995, **60** (1), 98-103.
14. C.S. James, 'Analytical Chemistry of Foods', Chapman and Hall, London, 1995, Chapter 5, p.104.
15. B.R. Bhandari, E.D. Dumoulin, H.M.J. Richard, I. Noleau and A.M. Lebert, *J. Food Sci.*, 1992, **57** (1), 217.

16. J.D. Joseph and R.G. Ackman, *J. AOAC International*, 1992, **75** (3), 488-506.
17. Anonymous, *Food Production*, Jan 1988, 17-20.
18. S. Anandaram and G.A. Reineccius, *Food Technology*, (1986), **40** (11), 88-93.
19. L.King, *Food Tech. Europe*, March/April 1996, 88-89.

INCORPORATION OF LONG CHAIN OMEGA-3 FATTY ACIDS INTO YELLOW FATS

S. Madsen

MD Foods
Skanderborgvej 277
PO Box 2470
DK- 8260 Viby J
Denmark

1 INTRODUCTION

MD Foods are one of the largest producers of yellow fats in Europe and functional foods is one of our focus areas. We have developed and introduced functional foods products over the last eight to 10 years; some of them yellow fats with incorporated long chain omega-3 fatty acids from fish oil.

In Europe as well as other advanced countries in the world some consumers are aware of the importance of eating fish. Nonetheless, in Denmark for example half of consumers never eat fish and in general in Europe the consumption of fish is below recommended levels. In contrast, the tendency is that the consumers are generally becoming more concerned about their eating habits.

2 BACKGROUND

The reason why MD Foods became interested in fish oil is based on the well-known studies performed by two Danish scientists, Bang and Dyerberg, who studied the disease patterns among 1,500 Greenland Eskimos in the years 1950-1970. In spite of the rather high intake of fat and cholesterol, the number of myocardial infarctions was significantly lower than in the Danish population (Table 1). The dietary fat of the Eskimos, however, came from fish, and Bang and Dyerberg were pioneers in connecting intake of omega-3 fatty acids with a reduced risk of myocardial infarction.

Many other studies have been performed in order to show the effect of fish oil on the reduction of deaths from cardiovascular diseases. Convincing data was given in a study performed by Professor M.L. Burr in which 2,033 men who had recovered from their first heart attack were divided into three groups.[1] Group 1 was advised to eat less fat in general, group 2 was advised to eat more cereal fibre and group 3 was advised to eat more oily fish. After two years the number of deaths was calculated in the three groups. Group 1 and 2 had the expected numbers of deaths. In group 3 the number of deaths was 29% lower than expected.

Table 1 *Disease Patterns Among 1,500 Greenland Eskimos in Upernarvik, 1950-1974*

Disease	Actual number of cases (1950-1974)	Expected number of cases*
Cancer (all forms)	46	53
Strokes (apoplexy)	25	15
Epilepsy (grand mal)	16	9
Peptic ulcer	19	29
Acute Myocardial Infarction	3	40
Psoriasis	2	40
Bronchial Asthma	1	25
Diabetes Mellitus	1	9
Multiple Sclerosis	0	2

*Based on the disease pattern in Denmark

3 OMEGA-3 INTAKE

How can consumers increase their dietary intake of Omega-3? The easiest way is to eat more oily fish. But as mentioned above half of the Danish population never eat fish. Fish is considered expensive and inconvenient.

Table 2 *Quality Parameters of Cod-liver Oils Purchased in the UK*

Product	Peroxide value	Anisidine value	Totox	Polymers	Oxidograph 100°C (CH)	Taste
UK-I	11.4	15.9	38.7	0.29	1.2	Unpleasant
UK-II	4.9	10.5	20.3	1.40	2.4	Less unpleasant
UK-III	3.0	14.7	20.7	0.85	3.3	Less unpleasant
UK-IV	3.9	58.2	66.0	0.88	1.8	Strong flavour
UK-V	2.0	11.3	15.3	0.27	3.3	Unpleasant
UK-VI	3.2	20.1	26.5	0.87	5.0	Unpleasant

Another way to increase intake is to eat capsules or cod liver oil. This results in the consumption of additional fat. Also capsules are expensive per gram of omega-3 fatty acids. Moreover, the fish oil in capsules is generally oxidised, giving a fishy flavour. A question mark can be put on the health value of oxidised fish oil (Table 2).

The third possibility is to eat food products enriched with Omega-3. MD Foods have developed the spread PACT to supply additonal Omega-3 fatty acids.

4 FISH OIL

If fish oil is to be used in food products a high quality must be maintained. During the whole production process focus must be put on avoiding oxidation of the fish oil. Some of the problems in the existing fish industry are:

- some of the fish is damaged before processing
- the fish has been stored for up to two weeks
- antioxidants, if any, were wrongly administered
- the refining temperature was too high
- the fish has been exposed to oxygen several times
- the fish has not been analyzed correctly
- fish oils are by-products of fish protein.

5 CASE STUDY—PACT

PACT is a fat reduced spread with unhydrogenated fish oil and it was marketed in England in August 1995. The aim was to produce the healthiest spread on the market by following the nutritional guidelines in the UK COMA (Committee on Medical Aspects of Food Policy) report *Nutritional Aspects of Cardiovascular Disease.*[2] One of the recommendations in this report was to use reduced fat spreads and to decrease the amounts of *trans* fatty acids in the diet. PACT is a reduced fat spread and is virtually free of *trans* fatty acids.

Other recommendations of the report were to cut down on saturated fat and no further increase in average omega-6 polyunsaturated fatty acids. PACT is low in saturated fatty acids, low in omega-6 polyunsaturated and high in monounsaturated fatty acids.

The COMA report also recommended a minimum intake of 0.2g omega-3 fatty acids a day and PACT provides 0.2g long chain omega-3 fatty acids in 15g product. 0.2g omega-3 fatty acids corresponds to 20-30g oily fish such as herring or 100g lean fish such as sole and cod. The average consumption of yellow fats in the UK is 30g/day.

Thus we had succeeded in developing a healthy spread, in line with the COMA report recommendations.

It was decided that 10 factory trials (1 every week) should be conducted to determine that the spread was free from fish oil off-flavour during a period of 16 weeks. The protocol was to test samples:

- stored at 7^0C

- stored at 7^0C but with 3 x 30 minute periods out of the refrigerator/week
- used in baking and frying tests.

The factory trials were unproblematic. Since the launch in August 1995 we have had no complaints about product quality.

6 CASE STUDY—BLUE GAIO

On the Danish market butter and dairy blends (Kaergarden -80% fat of which 75 parts butteroil/25 parts rapeseed oil) are dominant to margarine and low fat spreads. Some years ago MD Foods launched a reduced fat butter with only 40% butteroil, but the product never obtained an acceptable market share. The Danes prefer the butter taste/quality.

Blue Gaio is a dairy blend, produced using butter technology, enriched with long chain omega-3 fatty acids from fish oil. It was introduced to the Danish market in January 1997 to give the consumers a healthier alternative to butter and dairy blended products. One of the main objectives was to develop a product with consistency and taste as close to Kaergården as possible, so that consumers would accept the omega-3-enriched product as an healthy alternative. The sensory diagram below demonstrates how well this objective was achieved.

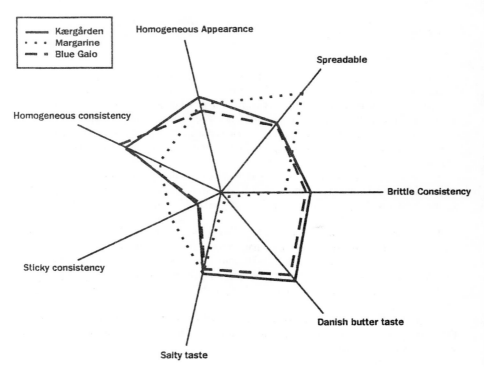

One of the reasons why MD Foods is successful with the products enriched with omega-3 fatty acids is that emphasis is put on ensuring the quality of the fish oil, in the following way:

- the ecological environment
- fresh edible fish
- optimal refining

The ecological environment: the fish are caught north of the Polar Circle where the Gulf Stream and the Arctic Ocean meet. There is no heavy industry and the ocean has a depth >1 km.

Fresh edible fish: The fish are edible, caught by a local coast fleet, which goes out in the morning and home in the evening. The fish are processed as soon as the fleet returns in the evening.

Optimal refining: This is achieved without the use of light, heat and oxygen. The fish oil is refined in a gentle way, shipped in special containers to Aarhus Oil and further on to the dairy. Several hundred handlings have been controlled when the fish oil finally is incorporated into the product.

7 REFERENCES

1. M.L. Burr, A.M. Fehily, J.F. Gilbert *et al. The Lancet* 1989, ii, 757.
2. Department of Health. Nutritional Aspects of Cardiovascular Disease: HMSO, 1994.

IV Regulatory and Consumer Issues

CONSUMER ISSUES AND FUNCTIONAL FOODS

J. Young

Leatherhead Food RA
Randalls Rd
Leatherhead
Surrey KT22 7RY

1 INTRODUCTION

As a way of introduction to this session on regulatory and consumer issues, I think it is particularly important to dispel any notion that the average UK or European consumer has any real understanding or indeed awareness of the term 'functional food'.

What we term functional foods are to the average consumer just ordinary everyday food and drink products carrying specific health claims. Realising this, it becomes important to establish whether consumer perception, purchasing patterns and attitudes relating to foods and drinks are influenced by some of the health claims now finding wider usage—eg 'can help lower cholesterol' or 'can help the body's natural defences.'

Before discussing this matter in more detail, I think it is important to recognise that widespread media attention, coupled with aggressive lobbying by consumer interest groups, has added to the number of consumer issues surrounding functional foods. Rather than just focusing on the central issue of whether claims are justified, the debate has broadened to include nutrition policy, product safety, pricing and consumer education, aspects of which will be covered later.

Before addressing the specific issues surrounding the consumer and functional foods, I think it is important to gain a broader understanding of the consumer's general perceptions and attitudes concerning health and wellbeing. I would like to do this by presenting some of the findings of a recent study by the Leatherhead Food Research Association, which examined the attitudes of British, French and German consumers to health and wellbeing.

2 THE EUROPEAN CONSUMER

Energy levels, physical appearance and absence of illness were cited by respondents as the three most important factors for judging their own state of health, although German respondents were the most likely to use blood pressure or cholesterol readings as a way of judging health. By far the most important reason for wanting to be healthy was simply to 'feel good'. It is of interest to note that concern for weight, once the driving force behind consumer interest in healthy eating, now appears to be somewhat lower down the list of consumer priorities.

Longevity was also thought to be an important factor behind consumer interest in health, cited particularly by the older age groups and those with children. Also worthy of mention is that 38% of UK respondents and 79% of German respondents cited the prevention of disease as a factor motivating desire to be healthy, compared with just 17% for the French.

Of particular significance to the concept of functional foods is that, when respondents were asked about the relative importance of diet, exercise and genetic make-up in contributing to health, diet was perceived as the most important factor overall, followed by exercise then genetic make-up.

Table 1 *Criteria on Which Respondents Judged own Health*

Base 605

	Percentage of Respondents		
	UK	**France**	**Germany**
Energy levels	75	65	69
Physical appearance	64	17	55
Absence of illness	62	34	77
Weight	51	15	39
Bowel movements	17	3	23
Blood pressure reading	17	4	32
Cholesterol reading	11	4	25

Source: Leatherhead Research Association

Table 2 *Relative Importance of Diet, Exercise and Genetic Make-up in Contributing to Health*

Base 605

	Percentage of Respondents		
	UK	**France**	**Germany**
Diet	68	55	70
Exercise	24	29	11
Genetic make-up	8	16	20

Source: Leatherhead Research Association

Differences between countries include the fact that French respondents put a higher emphasis on exercise than UK or German respondents, while UK respondents placed less emphasis overall on genetic make-up.

There is a fairly high degree of overlap in the medical conditions of most concern, with heart disease and stress at the top of the list.

German consumers appear to be more knowledgable about health matters than other European consumers. This is reflected in a concern about a broader range of conditions.

In the context of functional food development, it is important to note that the research indicated that raised cholesterol levels was seen as a condition over which the individual could exert a fairly high degree of influence, while cancer of the stomach or colon was not.

Table 3 *Consumers' Main Health Concerns*

Base 605

UK	France	Germany
Heart disease 55%	Stress 42%	Heart disease 49%
Stress 45%	Migraine 26%	Cancer of stomach 47%
High blood pressure 37%	Heart disease 23%	Stress 41%
Obesity 31%	Obesity 21%	Osteoporosis 38%
Cancer of stomach 29%	Memory loss 20%	High cholesterol 38%

Source: Leatherhead Food Research Association

Table 4 *Degree of Influence in Preventing Onset of Medical Conditions*

Base 605 — **Mean score out of 10***

Condition	UK	France	Germany
Heart disease	7.1	7.4	6.0
Stress	6.5	5.0	7.2
High blood pressure	6.9	7.1	7.3
Obesity	8.2	8.4	8.9
Cancer of stomach	4.6	6.1	3.9
Migraine	5.2	5.1	3.7
Memory decline	3.3	5.6	3.6
Osteoporosis	5.4	6.1	9.3
High cholesterol	7.7	8.1	8.7

*1=no influence and 10=considerable influence

Source: Leatherhead Food Research Association

The research implies that interest in health claims is likely to be high when associated with conditions that generate a high degree of health concern, eg heart disease, and where consumers feel they have a fair degree of influence.

Although there appears to be some overlap in the top five potential health benefits for the French, British and German consumers, the differences are sufficient to suggest that to be successful on a pan-European scale, a functional food strategy needs to take account of these national differences.

While the UK consumer places great emphasis on claims related to heart disease, ie lowers cholesterol, the German consumer is more interested in claims relating to general disease resistance.

It is also interesting to note that one of the health claims with which consumers were most preoccupied, 'prevents cancer', was the one over which they felt they could exert least control.

These results clearly demonstrate that the consumer is highly receptive to the concept of delivering health benefits through diet, a fact that food manufacturers have not been slow to pick up on. According to some research by Leatherhead Food Research Association in mid-1995, 67% of UK consumers thought functional foods were either a very or quite good idea.

However, the outcome of additional research by the Leatherhead Food Research Association indicates that a strong health claim alone does not guarantee success. As for all products, success rests on a combination of factors including quality, particularly flavour, consumer perception, product positioning, price, packaging, advertising, etc. Regarding functional foods, there is also the constant worry that any health claims will generate criticism from consumer pressure groups, thus invariably leading to adverse media reporting.

Table 5 *Most Popular Potential Health Claims*

UK	France	Germany
Reduces risk of heart disease 67%	Gives energy 42%	Increases resistance to disease 58%
Promotes healthy teeth/ bones 59%	Promotes healthy teeth/ bones 41%	Promotes immune system 56%
Prevents cancer 55%	Prevents cancer 35%	Promotes healthy teeth/bones 53%
Lowers cholesterol 45%	Reduces risk of heart disease /Aids memory function 26%	Improves athletic perfomance 52%
Lowers blood pressure 41%	Increases resistance to disease 26%	Promotes a healthy gut 46%

Source: Leatherhead Food Research Association

3 FUNCTIONAL FOODS IN THE UK

Indeed, functional foods have been notable in the UK more for their failure than their success. Why is this? Some pointers are given by a programme of qualitative consumer research undertaken by Leatherhead Food Research Association towards the end of 1995.

At the time of the research there was much speculation as to the extent of consumer interest in functional foods, with very little independent canvassing of public opinion as to whether there was a real need for them and whether products were meeting this need. The research was an attempt to address this issue by obtaining reactions of health-conscious individuals not only to the concept but also to actual products recently launched. It is this latter aspect that I would like to focus upon.

Table 6 *Positives and Negatives of Three Functional Food Products*

(I) Positives and Negatives of Ribena Juice & Fibre	
Positives	**Negatives**
+ Tastes good (particularly orange and apricot) + Attractive/practical packaging + Good for children with low fibre intake + Good for slimmers	- Too expensive - We get enough fibre already - Marketing ploy/fibre there anyway - Fibre suggests savoury/thick - Confusing soluble fibre message - Ribena bad for teeth - Fear of fibre overdosing (especially children)
(II) Positives and Negatives of Gaio	
Positives	**Negatives**
+ Advertising well remembered + Enjoyable, new taste + Attractive packaging + Made a change + Good for the heart/cholesterol	- Causido, unknown ingredient, term had unclear associations - Gaio confused with bio - Cholesterol claim confusing
(III) Positives and Negatives of Omega-3 Spreads	
Positives	**Negatives**
+ Good for the heart + Familiar packaging +Thought to have same benefits as cod liver oil + Replaces fish in the diet	- Unclear image created by the word 'omega' - Health claims not well understood - Might have fishy smell or taste - Oil suggests high content - Worries about how much omega-3 you would get

From this research I would identify four key reasons why functional foods have so far failed to capture the imagination of the UK consumer:

(i)   Some are too expensive
(ii)   Most of the health claims are poorly understood
(iii)   There are concerns about overdosing
(iv)   Some contain unfamiliar ingredients

This raises the question of the need for broader-based and more effective communication strategies.

A recent study by the UK National Consumer Council throws further light on the current level of consumer confusion surrounding health claims on food packs. The research, which was financed by the Ministry of Agriculture, Fisheries and Food, indicated for example, that consumers were confused by longer, more complex health claims and that many claims, lists and tables were meaningless and impenetrable.

This paper will now focus briefly on some of the broader consumer issues arising from the use of more strident health claims, some of which were aired during Leatherhead Food Research Association's programme of consumer research.

2.1 National Nutritional Policy

Clearly, there are concerns that a proliferation of health claims on food products could mask or confuse healthy eating messages, thereby leading to further consumer confusion.

2.2 Product Safety

Consumers appear to be worried about the possibility of overdosing on some of the new ingredients finding application in foods.

2.3 Pricing

There is the view that products offering potential health benefits should be priced so that they are accessible to the public at large.

2.4 Consumer Education

For functional foods to realise their full potential the consumer will need to be educated about their various health benefits, and how these foods should be used in the context of a normal dietary pattern. Based on the research findings of Leatherhead Food Research Association and the National Consumer Council this will not be an easy task.

Lack of knowledge, confusion and misunderstanding within the media, law enforcement agencies, the medical profession and particularly consumer groups also implies the need for more broadly based communication strategies.

4 HEALTH CLAIMS STRATEGY

The potential for adverse publicity creates a dilemma, as barriers to conveying health benefits associated with functional foods may well hold back consumer acceptance and thereby market development. However, closer examination of health claims strategies across Europe does indicate that it is possible to convey a health benefit to consumers, without running the risk of adverse media publicity or intervention by the regulatory authorities.

5 CONCLUSIONS

To conclude, this paper indicates that, while consumers are clearly receptive to the concept of improving their health and wellbeing through diet, the realisation of this is far from straightforward.

Clearly, for progress to be made new approaches to consumer education will be required coupled with a commitment from all interested parties to support any initiatives in this area. Indeed, the National Consumer Council study recommends further research on consumers' understanding of healthy eating messages and the part played by food packaging in targeting such information. They also recommend that MAFF should continue examining how best to target information and advice about diet.

If nothing else, I hope this presentation has served to highlight that health claims alone do not guarantee market success.

SYSTEMATIC REVIEW AS A METHOD FOR ASSESSING THE VALIDITY OF HEALTH CLAIMS

M. Rayner

British Heart Foundation Health Promotion Research Group
Division of Public Health and Primary Health Care
University of Oxford
Oxford OX2 6HE.

1 INTRODUCTION

This paper discusses what evidence a food manufacturer needs to make a health claim on a food packet or in an advertisement for a food and proposes that systematic review should be used as a method of critically evaluating the quantity and quality of that evidence.

Food manufacturers need to be able to inform consumers of the results of their research about the relationship between food and health. On the other hand consumers need to be protected against misleading health claims based on inadequate research.

There are now various diet-health relationships which are well established but there are many others which are less well established or which are controversial. Since food packets and advertisements are too small to explain the quantity and quality of the research on which a health claim is based there need to be rules about which claims might be made and which not.

Without such rules food manufacturers who use claims based on poor evidence will have an unfair competitive advantage over those who only use good evidence. Furthermore, consumers who attempt to use claims to help improve their health may be misled. The Ministry of Agriculture, Fisheries and Food has recently issued for consultation, draft guidelines for health claims which set out some general principles for making health claims.[1] One section of these guidelines deals with the substantiation of claims.

An example of where better methods for assessing the validity of health claims would have been helpful is the instance of the recent advertising for Gaio - a yogurt-style food produced by MD Foods. A promotional brochure claimed that *'Controlled clinical tests conducted by scientists in Denmark proved that the Causido culture [in Gaio]... can help lower the level of harmful cholesterol in the body.'* This claim was apparently based on a single study carried out by Agerbaek *et al.* on 58 44-year old Danish men.[2] Total cholesterol levels in the intervention group fell, over six weeks, from a mean of 6.07 to 5.71 mmol/L. No confidence limits for the reduction were given in the paper. A more recent study by Sessions et al involving 173 men and women (aged 30-55) failed to show any effect of consuming the product over 12 weeks.[3]

Was the paper by Agerbaek *et al.* sufficient evidence for MD Foods to claim that consuming Gaio can help lower cholesterol? If not why not?

2 SYSTEMATICITY AS A WAY OF REDUCING BIAS IN REVIEWS

Nutrition is just one area of medical science where there is a need to sort out the good evidence from the bad. Systematic review has become an established method for assessing the evidence for the effectiveness of medical interventions. Guidelines for systematic review have been drawn up by various bodies including the NHS Centre for Reviews and Dissemination in York[4] the Cochrane Collaboration[5] and others.[6]

The methodology for systematic review is still, to some extent, in development but in essence systematic review involves the following steps:

1.   Defining a precise research question or questions.
2.   Establishing inclusion/exclusion criteria for literature to be covered by the review. These inclusion/exclusion criteria should relate to:
     (i) the research questions of the review
     (ii) the study design of the primary studies.

3.   Systematic searching of the literature to ensure that all the available literature (both published and unpublished) which meets the inclusion criteria is located.
4.   Systematic data extraction from the literature—preferably by more than one reviewer.
5.   Data synthesis. Where possible this should involve meta-analysis—the use of statistical methods to summarise the results of independent studies.
6.   Unbiased interpretation of the data.

The best reviewers of the scientific or medical literature have always used systematic methods and there is no clear dividing line between a systematic review (as described above) and the best reviews of the past. But many reviews are not systematic to any significant degree and the process of carrying out reviews has often been a matter more of art than of science. (By reviews I mean not just reviews which describe themselves as such, but also introductions to papers, editorials, reports of expert committees, even text books.) It is increasingly realised that reviews, like primary research, need to be more scientific. Otherwise, like poorly conducted primary research, reviewing can be subject to bias.

Here is an example of a checklist which might be used to check whether a review is biased or not:

1.   Research question formulation
     Is the question clearly formulated?

2.   Study identification and selection
     Is the search for relevant studies thorough?
     Are the inclusion/exclusion criteria appropriate to the research question?
     Is the validity of included studies adequately assessed?

3.   Data synthesis
       How sensitive are the results to changes in the way the review is done?

4.   Interpretation of results
       Do conclusions exceed the evidence reviewed?
       Are conclusions linked to the strength of the evidence?
       Is 'no evidence of effect' interpreted as 'evidence of no effect?'

3 OTHER WAYS OF REDUCING BIAS IN REVIEWS

Before the recent interest in, and development of the methodology for, systematic review two methods had evolved for dealing with bias in reviews: expert consensus and peer review.

Governments, national and international health bodies frequently set up committees of experts, sometimes deliberately ensuring that a range of different views are represented, to come to a consensus about what the current research demonstrates. Having more than one person involved in a review may be better than having just one, and having 'experts' may be better than having 'non-experts' but expert committees are not necessarily free from bias in their interpretation of research evidence.

Expert committee reports do not normally describe the process by which the committee came to their conclusions so it is normally impossible to assess whether the literature has been reviewed systematically or not. For example, efforts may have been made to ensure that all the relevant literature was considered but the expert reports do not normally say whether this was the case and if so how.

Another method of seeking to ensure that reviews are less biased than they might otherwise be is peer review. Opinions differ over how this should best be carried out. Again peer review seeks to ensure that the conclusions drawn in a review are not solely those of the author but of one or more other experts in the field.

4 SYSTEMATIC REVIEW IN PRACTICE

A landmark example of systematic review with meta-analysis is illustrated in Figure 1. In the early 1980s there was considerable controversy regarding the benefits of streptokinase in the treatment of myocardial infarction, and as late as 1987 the Oxford Textbook of Medicine advised its readers that *'the benefits of thrombolysis... remain to be established.'*[7] Various small trials had been published but these seemed inconclusive. Some had shown positive and some negative effects.

In fact four years earlier Yusuf and his colleagues had published the results of a systematic review of 33 trials that compared intravenous streptokinase with a placebo or no therapy in patients who had been hospitalised for acute myocardial infarction.[8] Figure 1 shows the odds ratios and 95% confidence limits for the 33 trials. The left side of the figure shows that the effect of treatment with streptokinase on mortality was favourable in 25 of the 33 trials but only six gave a statistically significant result. The overall pooled estimate of treatment effect given at the bottom significantly favoured treatment.

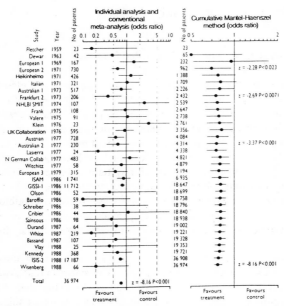

Figure 1 *Conventional and cumulative meta-analysis of 33 trials of intravenous streptokinase for acute myocardial infarction; Odds ratios and 95% confidence intervals for effect of treatment are shown* Source: Reference 6

The right side of the figure shows the same data presented as if a new or cumulative meta-analysis was performed each time the results of a new trial were reported. It shows that in 1971 the treatment effect became statistically significant for a two-sided P value of <0.05. This cumulative type of systematic review indicated that intravenous streptokinase could have been shown to be life saving almost 20 years before its submission to, and approval by, the US Food and Drug Administration and its general promotion in practice.[9]

An example of a systematic review of more direct relevance to the issue of health claims for foods is that recently published by Clarke *et al.*[10] These authors analysed the results of 395 dietary experiments relating to the effect of consuming different types of fat on blood cholesterol levels. Many of these studies were too small to be reliable, so selective emphasis on particular studies has previously led to conflicting conclusions about the results.

The meta-analysis of the results of all 395 studies demonstrates that isocaloric replacement of saturated fats by complex carbohydrates results in a fall in total blood cholesterol levels (by 0.52 mmol/L for a reduction of 10% in dietary calories from saturated fat), that similarly replacing complex carbohydrates with polyunsaturated fat leads to a fall in total cholesterol levels (by 0.13 mmol/L for an increase in 5% of calories from polyunsaturated fat), but that replacing carbohydrate with monounsaturated fat does not have a significant effect on total cholesterol levels.

5 THE MAIN AIMS OF SYSTEMATIC REVIEW

Systematic review has two main aims:
- firstly to ensure that all evidence relating to the research question is considered

• secondly to make certain that conclusions are drawn from the best quality evidence.

5.1 Assessing Methodological Quality of Included Studies

Although not sufficient for systematic review, assessing the quality of evidence is an important aspect of the process. The systematic reviewer needs to be both comprehensive and discriminating. The assessment of quality aims to grade studies according to the reliability of their results, so that they can be given appropriate weight in the synthesis of data and when drawing conclusions. This stems from the principal aim of reducing bias. Studies of high quality are likely to be least biased.

Those responsible for drawing up guidelines for systematic review have suggested that studies can be graded into a hierarchy according to their design, and that the hierarchy reflects the degree to which different study designs are susceptible to bias. Ideally a review will concentrate on studies which provide the strongest evidence, but where only a few good studies are available studies of weaker design may have to be considered.

Figure 2 shows one example of a hierarchy of evidence based on one drawn up by the NHS Centre for Reviews and Dissemination in York. This suggests that experimental studies are of higher quality than observational studies. Showing that differences observed between groups of subjects is the result of what you might think it is, is obviously much harder with observational studies than with experimental studies.

Figure 2 also suggests a hierarchy exists amongst experimental studies according to the exact mechanism which is used to allocate treatments. The most reliable method is generally considered to be random allocation. Observational studies can be graded according to whether they are prospective or retrospective, etc.

I	Well-designed randomised controlled trials
	Other types of trial
II-1a	Well-designed controlled trial with pseudo-randomisation
II-1b	Well-designed controlled trial with no randomisation
	Cohort studies
II-2a	Well-designed cohort (prospective study) with concurrent controls
II-2b	Well-designed cohort (prospective study) with historical controls
II-2c	Well-designed cohort (retrospective study) with concurrent controls
II-3	Well-designed case-control (retrosprective study)
III	Large differences between times and/or places with and without intervention
IV	Descriptive studies

Figure 2 *Example of a hierarchy of evidence*
Source: Adapted from Reference 4

5.2 Ensuring that All Evidence is Considered

With the development of electronic methods of literature searching it has become easier to ensure that all, or at least the majority, of the literature relevant to a review is considered. Much has been written on the best ways of searching the published literature using computerised databases of the literature.[11]

Another problem with traditional reviews that systematic reviewers are beginning to develop methods of dealing with, is publication bias, ie the tendency for positive studies to be published and for negative studies not to be.

The possibility of publication bias can be investigated during the data analysis phase of a systematic review.[12] A plot of the distribution of effect sizes according to sample size should make a funnel shape—there being more variability in reported effect sizes for smaller studies. Large gaps in the funnel indicate a likelihood of publication bias. An example of two funnel plots one showing evidence of publication bias and the other not is shown in Figure 3.

Publication bias in a diet-health area has been clearly demonstrated in a meta-analysis of the effect of garlic in the treatment of moderate hyperlipidaemia. A funnel plot of the published studies of the effect of garlic on blood cholesterol levels shows that it is likely that many small studies showing a negative effect of garlic have not been published.[13] A large 'mega-trial' of the effect of garlic on blood cholesterol levels is now needed because the evidence for publication bias means that the published evidence cannot resolve whether there is an effect or not.

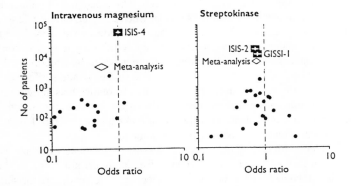

Figure 3 *Funnel plots for meta-analyses refuted and confirmed by subsequent mega trials: intravenous magnesium (left) and streptokinase (right) in acute myocardial infarction. Points indicate odds ratios from small and medium sized trials, diamonds indicate combined odd ratios with 95% confidence intervals from meta-analysis of these trials, and squares indicate odd ratios with 95% confidence intervals from mega trials*

Source: Reference 12

6 APPLYING SYSTEMATIC REVIEW TO ASSESSING THE VALIDITY OF HEALTH CLAIMS

It is generally assumed that health claims should be based on good scientific evidence but only rarely is it explicitly stated what this means in practice. Because of an apparent increase in health claims for foods, guidelines or rules for their use have been, or are being, drawn up in many countries. But in developing these guidelines and rules the question of what is meant by good scientific evidence has only rarely been considered.

6.1 The US Nutrition Labelling and Education Act of 1990

This states[14] that:

> *'the FDA will promugalte regulations authorising a health claim only when it determines, bsed on the totality of the publcy available scientific evidence (including evidence from well-designed studies conducted in a manner which is consistent with generally recognised scientific procedures and principles), that there is significant agreement among experts qualified by scientific training and experience to evalute such claims, that the claims is supported by such evidence.'*

In proposing that health claims should be based on the 'totality' of the published literature and not just a selection, the US Nutrition Labelling and Education Act does at least ensure some systematicity in the review process. But note that the Food and Drug Administration are only required to examine published literature and need pay no regard to unpublished evidence. As we have seen, taking account of publication bias and searching for unpublished literature is crucial when carrying out a review.

The Act also indicates that the evidence should include well-designed studies but gives no precise guidance about what level of quality of study design is considered sufficient. For example the Act is unclear whether there needs to be experimental studies before a claim can be made or whether observational studies would suffice.

The Act requires the FDA to use expert consensus to reduce bias in their review of the evidence. It indicates that there needs to be *'significant scientific agreement'* amongst experts that the evidence is sufficient to make a claim. Here I suggest that expert consensus as a means of reducing bias is inadequate.

A major problem with relying on expert consensus, for example when assessing the validity of health claims, is that expert reports frequently do not come to the same conclusion. There then has to be some means of selecting which expert reports are most reliable.

The FDA has now authorised the use of a number of health claims and disallowed others. The authorisation process is extensively documented in the Federal Register. The Register shows, for example why claims relating to saturated fat consumption and risk of coronary heart disease (CHD) were authorised and claims relating to n-3 fatty acid intake and risk of CHD were not.[15]

When it came to deciding whether to authorise claims relating to n-3 fatty acid intake and CHD risk the FDA basically relied on two expert reviews: the National Research Council's 1989 report *Diet and Health: Implications for reducing chronic disease risk*[16] and the Surgeon General's 1988 report *Nutrition and Health*.[17]   Both of these came to the conclusion that there was insufficient evidence of a relationship. The FDA rejected the findings of other expert reports such as that of the Life Sciences Research Office of the Federation of American Societies of Experimental Biology[18] that there was enough evidence.   The FDA did not appear to have any systematic way of determining the reliability of expert reports.

6.2 The Draft Guidelines on Health Claims Recently Issued by the Ministry of Agriculture Fisheries and Food in the UK

These state[1] that:

> '*Any health claim should be supported by a dossier of scientific evidence demonstrating the specific physiological effect which is claimed...*

> '*The scientific evidence for the health claim should be based on studies in humans and should include epidemiological evidence.   Any epidemiological studies need to:*
> *- be carried out in a representative cross-section of a population similar to that of the UK or in a representative sample of the sector of the population at which the product is aimed;*
> *- involve the study group consuming a reasonable quantity...of the food at a reasonable frequency...;*
> *- be of sufficient duration to ensure that the beneficial effect is maintained over a reasonable period of time...*
> *- take into account confounding factors...*
> *- produce statistically significant results.*

> '*The dossier in support of a health claim should be available for scientific peer review....*'

Under the UK draft guidelines, the review of the evidence on which a claim could be based (the dossier), might be even less systematic, than under the US Nutrition Labelling and Education Act.

In contrast to the US Act, the draft guidelines make no mention of the need to base a claim upon the totality of the evidence. The draft guidelines make no mention of how the proposed dossier of evidence is to be compiled. Theoretically there would be nothing to stop a food manufacturer selecting only those studies which substantiated the claim.

The guidelines give no guidance on the number or size of studies which need to have been carried out to support the claim. The Gaio case shows the problems of basing claims on single studies.

The draft guidelines do make some reference to the methodological quality of the studies on which the claim is based and recognise that the results of some studies might be the result of confounding factors. But the guidelines make no mention of how confounding is to be avoided (eg by the use of randomised controls).

It should be possible to draw up more detailed guidelines on the quantity and quality of evidence required before making a health claim. These guidelines could at least specify that the review of the evidence on which the claim is based must be carried out systematically.

7 CONCLUSIONS

While mainstream medical science has become increasingly sophisticated in its methods of reviewing the scientific evidence on which to base clinical decisions, those concerned about the validity of health claims have seemingly ignored these developments.

The US system for assessing the validity of health claims is better than the proposed UK system but both have major flaws. The US system relies too much on expert consensus. The UK guidelines would not ensure that the totality of the evidence is considered.

Any review of the evidence on which a health claim is to be based should be systematic ensuring that all the evidence is considered and that the studies included meet defined standards of methodological quality.

8 REFERENCES

1. Ministry of Agriculture Fisheries and Food. 'Draft Guidelines on Health Claims on Foodstuffs', 1996.
2. M. Agerbaek, L.U. Gerdes and B, Richelson, *Eur. J. Clin.Nut.*, 1995, **49**, 346.
3. V.A. Sessions, J.A. Lovegrove, T.S. Dean, C.M. Williams, T.A.B. Sanders, I.A. Macdonald and A.M. Salter [this volume]
4. NHS Centre for Reviews and Dissemination. 'Undertaking Systematic Reviews of Research on Effectiveness. CRD Guidelines for Those Carrying Out or Comissining Reviews', CRD, York, 1996.
5. Cochrane Collaboration, 'The Cochrane Collaboration Handbook', Cochrane Collaboration, Oxford, 1997.
6. I. Chalmers and D.G. Altman (eds), 'Systematic Review', British Medical Journal Publishing Group, London, 1995.
7. B.L. Pentecost, in D.J. Weatherall, J.G.G. Ledingham and D.A. Warrell (eds), 'Oxford Textbook of Medicine. Vol2. 2nd ed', Oxford University Press, Oxford, 1987, p.13.
8. S. Yusuf and P. Sleight, *Drugs*, 1983, **25**, 441.
9. J. Lau, E.M. Lantman, J. Jimenez-Silva, B. Kupelnick, F. Mosteller and T.C. Chalmers, *N. Engl. J. Med.*, 1992, **327**, 248.
10 R. Clarke, C. Frost, R. Collins, P. Appleby and R. Peto, *B.M. J.*, 1997, **314**, 112.
11 K. Dickersin, R. Scherer and C. Lefebvre, in I. Chalmers and D.G. Altman (eds). 'Systematic Review', British Medical Journal Publishing Group, London, 1995, p.17.
12 M. Egger and G. Davey Smith, *B.M.J.*, 1995, **310**, 752.
13 H.A.W. Neil, C.A. Silagy, T.Lancaster, J. Hodgeman, K. Vos, J.W. Moore, L. Jones, J. Cahill and G.H. Fowler, *J. Roy Coll. Phys.*, 1996, **30**, 329.
14 Food and Drug Adminstration, *Federal Register*, 1993, **58**, 3; 2478.

15 Food and Drug Administration. *Federal Register*, 1993, **58**, 3, 2682.
16 National Research Council/National Academy of Sciences, 'Diet and Health, Implications for Reducing Chronic Disease Risk', National Academic Press, Washington, 1989.

THE FUTURE OF FUNCTIONAL FOODS

J.T. Winkler
Food & Health Research
28 Paul Street
London
N1 7AB

1 INTRODUCTION

There is a gross disjuncture between the rhetoric and the reality of functional foods. Throughout the 1990s, conferences, reports, books and articles on the subject, both scientific and commercial, have proliferated. Most of the words written and spoken on functional foods have been enthusiastic about them, both for their contribution to public health and as profitable products. Yet in the shops of Britain, the number of foods that might plausibly be called 'functional' is modest.

Most of the products which have appeared have been recognisably just fortified foods. They have incorporated a broader range of fortificants, beyond the traditional vitamins and minerals. They promise some unusual if imprecise health benefits, like strengthening the immune system. But the concept is familiar, not revolutionary.

At the level of the individual product there is a wide gap between strong claims and weak sales. In the past three years 21 straight complaints against functional products have been upheld by the UK Advertising Standards Authority (ASA), because the claims made for the products lacked adequate substantiating evidence.[1] Despite such loose promotional techniques, no functional foods in Britain have been conspicuously successful. Two of the pioneer products which attracted attention on launch, Ribena Juice & Fibre and Gaio yoghurt, have already failed and been withdrawn.

Not for the first time, marketing hype outruns scientific evidence, inviting the conclusion that the whole concept of functional foods has been over-promoted, just another food fad for the high-tech era. Such scepticism affects policy makers as well as consumers. The Food Advisory Committee (FAC), for example, opens its review on the subject, not by anticipating growth, but by noting that '*The market for functional foods in the UK is at present very small....*'[2]

It is important, therefore, to look beyond short-term commercial prospects to consider long-term scientific prognosis. From that perspective, the importance of functional foods is being, if anything, underestimated.

Substantial research—governmental, academic and corporate—is in progress in the underlying sciences, increasing understanding of the relationship between diet and health.

This could provide the foundations for the functional foods of the future, more numerous and more significant than those which have appeared so far.

At least, that is the potential. What actually happens will depend in large measure on how we manage and apply this scientific research—that is, on the policy frameworks, public and private, within which functional foods are developed.

2 SCIENTIFIC FOUNDATIONS

Three fields are particularly relevant for the development of new functional foods:
(1) nutrition science and cell biology
(2) genetics
(3) food science and molecular biology.

2.1 Nutrition Science and Cell Biology

Even optimists acknowledge that we have had a *'poor understanding of the mechanisms by which food components affect physiological functions'*.[3] But as Professor Righelato, Director of the Functional Foods Initiative for the Ministry of Agriculture, Fisheries and Food (MAFF), concluded *'Nutrition has moved from being largely epidemiology based to understanding the physiological and genetic mechanisms by which diet and individual food components influence health and disease'*.[4]

Thus, we now begin to understand the *'mechanisms underlying the physiological, metabolic and behavioural effects of specific components of the diet'*.[3] We are charting metabolic pathways in greater detail, for a larger number of food components, and assessing their consequences for health, both positive and negative.

2.2 Genetics

From experience alone, we have long recognised that diet-related diseases run in families. At an elementary level, we have known that the susceptibility to these problems was inherited. Geneticists, working as part of the Human Genome Project and independently, are now mapping the genetic pre-disposition to diet-related diseases by showing how *'individual genetic make-up affects the response to dietary components'*.[3] For example, some people metabolise fat less efficiently than others and hence are more prone to coronary heart disease. Others process salt better than others and so are less liable to strokes.

Advances in genetics have implications for both individuals and public health. They are already leading to genetic screening of individuals for disease risk, in selected areas like breast cancer and hypertension, and the range of detectable conditions is likely to expand. In time, we will be able to assess the genetic component, if any, in the aetiology of the full range of diet-related diseases.

This will make possible more individualised dietary recommendations. Governments have previously emphasised that quantified dietary advice should be seen as desirable averages for the whole population *'not as recommendations for intakes by individuals or groups'*, as the Committee on Medical Aspects of Food Policy (COMA) advised in its most recent guidance for Britain.[5] In future, we may anticipate more personalised advice for individuals with specific genetic characteristics.

Once we reach that stage, industry will market specific products for consumers with specific needs. In the view of the Chairman of COMA, Professor Whitehead, '*Food products will be much more precisely targetted at sub-groups of the population*'.[6]

The concept of targetted foods is already familiar, for example, in diabetic products. But it will spread more widely, both in the number of foods and the number of at-risk groups catered for. As such products proliferate, they will blur the distinction between normal foods and clinical nutrition products or medical foods.[7]

2.3 Food Science and Molecular Biology

What is it exactly in fruit and vegetables which produces their demonstrated benefits for health? To answer that question in part, one strand of research now underway is focusing on phytochemicals, the thousands of other components in foods beyond the established nutrients. In parallel, herbalists of both the western and eastern traditions are seeking to demonstrate by standard scientific methods the efficacy of long-established remedies.[8]

We may anticipate that, in time, some of these other bioactive substances will prove to have beneficial consequences for human health. Current subjects of particular interest include, for example, the carotenoids and flavonoids. They will be extracted, refined and adapted into commercially useable forms, as functional ingredients for manufactured foods, as dietary supplements, or both. This will lead to '*compositionally enhanced foods to protect against degenerative diseases*'.[3]

At the same time, we are also increasing our understanding of the '*genetic basis of (the) nutritional characteristics of plant and animal products*'.[3] '*Scientific advances, including those in molecular genetics, now make possible a more rational and systematic modification of raw material*'.[3]

For example, one current controversy concerns soybeans genetically modified for herbicide resistance. But the next phase of research focuses on modifications to enhance their nutritional properties, particularly increased isoflavone content.[9] If this and similar research is successful, we may expect to see functional products with their '*quality enhanced at primary production rather than (through) processing*'.[3]

Functional foods of the future are thus likely to incorporate a vastly increased range of functional ingredients. They will not just be added to products as fortificants, but some will be built into the raw materials.

All quotations in this paper so far (except those identified by name) have intentionally been drawn from a single source, the 1997 Food Directorate Strategy of the Biotechnology and Biological Sciences Research Council (BBSRC). This is not the only or even the best source for evidence on these themes.[10,11] It was chosen because the strategy articulates the assessments of a government research council. These are fields of investigation with enough substance and promise to merit substantial state funding and co-ordination.

The developments summarised above, therefore, are not technologists' dreams or science fiction. They are mainstream, orthodox science. They are already underway and BBSRC support will ensure their medium-term expansion. They are the scientific foundations for the functional foods of the future.

To acknowlege this activity is not to replicate the over-enthusiastic futurology noted at the beginning. Potential problems abound.[12] Some of the research will fail. Some products brought to market will not prove effective. Others will be too expensive for the poor who need them most, and hence will make no contribution to public health. Some useful foods will not be commercially successful. Others will be too successful, creating new excess intakes and distorting the total diet. Unanticipated negative consequences will emerge, as with the use of genetic screening to exclude people from insurance or employment.[13,14]

New, scientifically-grounded functional foods will create competitive pressures on rival companies to offer copycat products, without doing the research. In future, as now, products will claim virtues they do not possess. Without a revised framework of law to take account of the new science, exaggerated, misleading and unsubstantiated claims would likely increase and become more serious—not just the routine improbable hyperbole of advertising ('best taste ever'), but false promises of improved health to anxious people ('worried-well' marketing[7]). Indeed, continued irresponsibility in promotion could discredit the whole category of functional foods, rather like patent medicines at the end of the 19th century.[15]

Undoubtedly, some of the current scientific effort will yield effective results —sufficient probably to effect a fundamental change in our understanding of the relation between food and health. We will be able to specify, with radically increased precision, which components in which foods produce which effects on which people—and to act constructively on that new understanding.

3 FOODS OF THE FUTURE

Imminent scientific developments are also likely to have a major impact on the food industry, substantially increasing the volume and variety of functional foods. From research already in progress, we may anticipate some of the categories of manufactured products which will be available next century.

3.1 Enhanced Variants of Traditional Foods

Products fortified with functional ingredients will continue. But they will increasingly be supplemented and replaced by products without anything 'added'. Some enhancement will be achieved by incorporating raw materials with genetically modified nutritional profiles. In other cases, different processing techniques will be used to increase the bioavailability of the beneficial components already present, as for example with toasting soya or microencapsulating n-3 polyunsaturated fatty acids.[16]

3.2 Disease-Specific Foods

Much recent innovation in the food industry has focussed on processed products which present less risk of diet-related diseases, because they are formulated with lower levels of fat, sugar or salt. The promise of functional foods is grander—the delivery of positive health benefits.

These potential gains include, explicitly, the prevention, treatment and cure of specific diseases. This possibility has not featured in the current policy debates about functional foods out of deference to the long-standing prohibition, in Britain and elsewhere, on any food claiming to have such 'medicinal' effects.[17]   But again, it is important to distinguish marketing and scientific issues.

Scientifically, there is no doubt that nutrients, and the foods which contain them, do have prophylactic and therapeutic effects. This is, after all, their function in the classic nutritional deficiency diseases. For example, the COMA report on Dietary Reference Values declares that the '*essential and undisputed roles of vitamin C (ascorbic acid) in man are to prevent scurvy...(and)...10 mg/d (is) also sufficient to cure clinical signs of scurvy*'.[5] Some vitamin C capsules are indeed licenced as medicines.

A currently topical example, which also points towards the future, is folic acid. Its role in the prevention of neural tube defects is now accepted.[18]   And its effect in lowering homocysteine levels may in time be generally recognised as giving folic acid a preventive role for coronary heart disease—moving beyond a traditional deficiency problem to one of the contemporary nutritional diseases of excess.

We may anticipate that eventually other bioactive food components will also prove, by established scientific standards, beneficial health effects.They will then be incorporated as functional ingredients into disease-specific functional foods.

3.3 Risk Group Specific Foods

There are gradations of risk for diet-related diseases and corresponding variations in the extent to which specialist foods are produced and consumed. At the hard end, where a condition has already been diagnosed, as with coeliacs and allergy sufferers, clearly identified products are widely available, for example, gluten-free and nut-free.

Where the risk is relative rather than absolute, specialist products are less common. For example, sugar-free confectionery targetted on children, the group most at risk of tooth decay, has been moderately successful in Switzerland, but hardly exists in the UK outside chewing gum.

The coming widespread use of genetic screening, which identifies an ineradicable risk to lethal, diet-related diseases, is likely to provide a significant stimulus to the development of specialist products for at-risk groups— functional foods for vulnerable people.

Less certain is how far specialist products will emerge for the lower end of the risk continuum.   We already have some foods for groups at a few nutritionally sensitive life stages—infants and pregnant women, for example. The European Union's PARNUTS directive, covering 'foods for particular nutritional uses', provides a legal framework in which such products might be developed, if the concept of functional foods becomes firmly established.

3.4 Foods to Reduce the Effects of Aging

The older population is one lifestage group for which specialist products may be confidently predicted.The idea that dietary modification can alter genetically-influenced life span and rates of aging has a long tradition in research on many animal species.[19] But the best-documented approach, caloric restriction,[20] is unlikely to ever prove popular with either consumers or the food industry.

However, the demographic shift in most developed societies towards older populations has put pressure on national health care budgets, finally provoking serious interest in preventive health policies,[21] including nutritional prevention of diet-related diseases.[22] These changes coincide in time, not only with the scientific advances described earlier, but also with an unprecedented widespread popular concern about diet.

This confluence of social, economic and scientific trends is stimulating specific research on dietary interventions to reduce the rate of aging[23] and to ameliorate the non-fatal diseases of aging.[24] The currently most advanced research focuses on lipid balance and antioxidants, especially vitamin E.

But a more general, theoretical, view sees maintaining the best possible nutritional status in the face of biological decline as the means to preventing premature disease and debilitation,[24] including cognitive degeneration. *'Many conditions associated with aging, such as loss of appetite and forgetfulness, may be avoided if optimal nutrition is maintained'.*[25]

Together, these developments also create commercial opportunities for specialised food products targetted at the elderly. We can see the beginnings of this trend already, with the opportunistic surge in Japan of products fortified with docosahexaenoic acid (DHA), claiming to retard senility, or the launching in Britain of a mixed fatty acid capsule alleged to 'keep old age at bay'.

A more complex indicator of the future is Campbell's Intelligent Cuisine range, now on trial in the USA—complete diets of popular meals, nutritionally modified for relevance to degenerative diseases, and designed to take account of the particular limitations of older consumers, hence pre-prepared and delivered direct to homes in bulk.[26]

One prominent component in the debate on nutrition and aging has broader relevance for the future of functional foods—the concept of 'optimal nutrition'. The general idea is *'to help improve physiological function of the individual by matching unique biochemical needs with nutrient intake'.*[24] These needs are determined by, among other factors, genetic inheritance and lifestage.

The immediate practical manifestation of this approach involves possible revision of the recommended daily intakes for vitamins and minerals, above levels of 'sufficiency' to some definition of 'optimality'. This is already explicit in UK government advice on folic acid[18] and more general application is currently under consideration by, among others, MAFF's Optimum Nutrition Programme. Manufacturers of diet supplements have been pressing this argument for some time. But the logic applies equally to fortified foods and the other bioactive substances or functional ingredients which they contain.

3.5 Performance Foods—Physical

The idea of constructing an optimal diet to improve physical performance is the foundation stone of animal nutrition—not just in the commercial livestock and poultry industries, but also in many other contexts where people regularly feed animals, as with race horses, domestic pets and wild birds. 'Optimal' includes not just concepts of balance, but of enhanced doses of specific nutrients, incorporated into fortified feeds.

The same principles apply to human animals, and are applied to them as well. They are not usually discussed publicly with equivalent explicitness or enthusiasm, perhaps out of sensitivities about 'social engineering'. The major exception in normal human nutrition is child development; concepts of an optimal diet are embodied in law, through regulations on the composition of infant formulae, drinks and foods.[27]

The idea of optimal nutrition for physical performance is central to the growing field of sports nutrition. Such concerns are no longer focused only on high-performance athletes, but are achieving popular recognition through a variety of sports drinks and foods promising, among other things, instant energy, stamina, optimal hydration, electrolyte balance, rapid recovery from fatigue, restoration of glycogen stores, muscle bulking and weight gain. Throughout the 1990s such products have achieved increasingly wide distribution, no longer just in specialist sports shops, but for example, in own label ranges of national retailers like Boots and in mass-marketed soft drinks.

The arrival of popular aging foods and 'megadose' vitamin and mineral supplements will further disseminate the concept of optimal nutrition for enhancing or at least maintaining physical capabilities. In future, we may anticipate many other functional foods claiming to optimise something.

3.6 Performance Foods—Mental

What is optimised may be mental as well as physical performance. The gains in alertness produced by caffeine are well-established, both scientifically and in popular lore. Claims to improve short-term memory through choline and gingko biloba are the subject of recurrent challenges[28] and the products containing them are still fringe items available only through specialist outlets.

With long-term mental development, the benefits of DHA are now accepted[29] and it is increasingly being added to infant formulae, initially in the UK by Milupa. The run on vitamin supplements, following media reports about their apparent enhancement of IQ, suggests that the commercial implications of research in this area may be large. A possible harbinger of the future is the success of a DHA-fortified soft drink in Japan, targetted on adolescents facing school examinations, marketed by Coca-Cola.

The greatest impact, however, both in terms of public health and private profit, may lie in nutritional approaches to retarding Alzheimer's disease.[25,30] It now seems probable that functional products for this purpose will appear early next century.

3.7 Mood Foods

Foods for treats, celebrations and rewards, providing gratification, compensation and indulgence, have long been staples in the marketing of sweet and snack products. There is now underway a serious research effort on ingredients and foods to alter psychological states.

The most advanced and significant strand of this work, from a public health perspective, concerns satiety. Such research has been reinvigorated by the recent rapid increases, in Britain and other developed countries, of obesity and overweight. The rationale is to manipulate perceived repletion as a means to appetite control and calorie reduction, with particular research comparing protein and carbohydrate-based meals.[31]

Tryptophan and serotonin feature prominently in satiety research and also in work on foods to relieve stress and induce calm. Indeed, serotonin has already been advanced, unsuccessfully, to support the claim that 'a Mars a day' helps you rest as well as work and play.[32] And confectionery manufacturers generally are prominent among those driving research in this area.[33]

At the opposite end of the mood continuum, capsules and drinks containing ephedrine and ma huang are already on the market, promising a 'legal high'. Predictably, they have proved popular as well as dangerous and controversial.[34] This may be the beginning of a major extension in the idea of functional products. There are certainly temptations for everyone to find a feel good factor in a food.

4 PRACTICAL IMPLICATIONS

If the predictions in this paper prove to be true, they would have many implications for everyone in the food chain, more than can be considered here. But to illustrate the point, let us examine some consequences for three topical issues—substantiation, claims and the concept of functional foods.

4.1 Substantiation

The two most fundamental questions about any functional food are: is it safe? does it work? Derivatively, two more questions are: how much and what kind of evidence is needed to answer those questions?

The new EU directive on novel foods[35] makes safety assessments mandatory. But the key question is: what is 'novel'? What could become a major loophole to evade testing lies in the provision that marketeers may declare new products 'substantially equivalent' to existing foods. In any case, there is no requirement to prove efficacy. In Britain, the FAC draft guidelines on functional foods[2] contain only limited and voluntary proposals on efficacy.

The promise of functional foods is to improve the health of consumers, in some degree analogous to medicines. As the current research yields new products with stronger and more specific effects, functional foods will approximate closer to drugs. In consequence, we must anticipate rising expectations that functional foods will have to meet higher standards of proof for efficacy, with correspondingly more rigorous assessment of their contraindications.

In medicine, since Cochrane exposed the spurious therapeutic claims of numerous drugs and procedures,[36] randomised controlled testing of all innovations for their effects on final health outcomes with human subjects has become the routine requirement. Normal practice in the food industry at present is well below this standard. But the unexpected negative results of two long-term, large-scale clinical trials of beta-carotene supplementation emphasize their value.

A sudden requirement of similar trials in the food sector is improbable. But over time, as research findings increase and the market develops, we may anticipate an incremental raising of the scientific requirements demonstrating efficacy for functional foods.

4.2 Claims

The present framework of law for food claims is simultaneously too strong and too weak—prohibiting some important legitimate claims, while *de facto* tolerating others which are misleading and unsubstantiated.

On the one hand, it was noted earlier that many nutrients and the foods which contain them have prophylactic and therapeutic effects. But virtually everywhere such medicinal claims on behalf of foods are forbidden. The research now underway is likely to prove more such beneficial effects, and hence sharpen this conflict between science and law.

Conversely, the EU has failed for almost two decades to produce a general directive on food claims. In this stalemate, member states recycle incomplete and out-dated regulations, left over from an earlier era. In Britain, despite greatly increased public interest in nutrition, the claims controls in the 1996 Food Labelling Regulations are virtually unchanged from those of 1984.[17]

Against this perverse legal background, marketeers of functional foods have responded pragmatically, with suggestive neologisms—for example, that products maintain health rather than prevent disease, reduce risk factors rather than the disease itself, provide protection rather than treatment, etc. The consequence is a proliferation of impressive but imprecise claims, which often confuse consumers.[37] The sequence of successful complaints against unsubstantiated claims[1] highlights the problem, but represents only a small fraction of contemporary practice.

The current intense phase of research will provide further stimulus to functional claims, both genuine and spurious. Companies will naturally promote successful developments. Rivals will respond, in a context which provokes and facilitates ambiguous hyperbole. Many consumers have anxieties about their health or aspirations for social success. They are ignorant of the nutritional basis underlying functional foods, and are susceptible to pseudo-science. The current legal framework provides few restraints.

In future, we may see two contrasting approaches to claims control—differentiation and simplification. At present, several regulatory authorities around the world are developing complex taxonomies of claims, each with their own rules for substantiation and usage.[38] Suggested claim types, not always clearly distinguished from one another, include: nutrition, nutrient, health, medicinal, therapeutic, generic, specific, risk factor, health maintenance, descriptive, user, diet, dietary recommendation, nutritional function, physiological function, nutrition messages, statements of nutritional support, foods for special dietary uses, specified health uses, and particular nutritional uses. Even if such frameworks reach the statute book, they may have limited impact on the real world, because of the difficulty of interpreting and administering them consistently.

The alternative is a radically simpler system in which there are only two types of claims—content claims, about what is in the food, and effect claims, about what the food does to the consumer. The operating principle is to authorise whatever claims science substantiates and establish effective controls over the rest.

In a Europe where impasse has prevailed for so long, it would be mere speculation to predict what, if indeed any, new regulatory framework will emerge. But one specific change seems probable. Sooner or later, under the pressure of mounting research findings, the illogical prohibition on medicinal claims will be removed. The *quid pro quo* will be much more demanding requirements for their scientific substantiation.

4.3 The Concept of Functional Foods

'Functional foods' is a marketing term, coined in Japan in the 1980s. It has a vivid, contemporary feel which captures the essence of a phenomenon, but does so imprecisely. The inadequacies of the term are a recurrent theme in the literature. Many authors propose their own definitions, so varying concepts proliferate. Goldberg uses three in the preface to his book alone.[7] Some themes appear frequently. A composite definition, built up from the most common, might read 'functional foods are normal foods for healthy people, with something added, providing a benefit beyond nutritional effects'. If the analysis presented in this paper is correct, then every component in that definition is wrong.

4.3.1 Normal. Some functional foods will be disease specific, creating a continuum that extends into clinical nutrition products. Most commentators explicitly exclude such therapeutic foods. Wrick is unusual—and I believe right—for including them and analysing their significance, for marketing as well as health.[7]

4.3.2 Foods. Many writers exclude products presented as pills or powders, resembling drugs. The form of the product becomes a component of the definition. But the critical part of any functional product is the functional ingredient, the bioactive substance which produces the health benefit. These are already sold in multiple formats, as with vitamin C, or aloe vera, about which the ASA upheld a complaint against a product marketed in three formats simultaneously. This pattern is likely to be commonplace with the functional phytochemicals of the future.

4.3.3 Healthy. Functional products will be marketed for consumers at varying degrees of risk, for older people at various stages of degenerative diseases, for the already ill seeking to avoid relapse, and for the genetically compromised, as well as the apparently well.

4.3.4 Added. Fortified and enriched products will increasingly be supplemented and replaced with 'no added' functional foods, made from genetically modified raw materials or processed to enhance the bioavailability of the functional ingredients already present.

4.3.5 Beyond Nutritional Effects. Concepts of what constitutes a nutritional contribution, or indeed what constitutes a nutrient, are changing. Even with the conventional micronutrients (vitamins and minerals) we are now considering sufficient, optimal and therapeutic doses. With the functional ingredients of the future, we will recognise a spectrum of effects, without trying to divide them into nutritional and other effects.

The retrospective attempt by scientists to inject some rationality into marketing jargon will prove increasingly inadequate. In time, the term 'functional foods' will lose its connotations of futuristic trendiness and may be discarded.But the perspective of functionality will continue and strengthen—analysing foods and ingredients in terms of the functions they perform for those who eat them.

In the next century, accumulating research results will substantiate in detail the health effects of an expanding range of bioactive components in foods, giving new relevance to one of the clichés of this field—all foods are functional. But if the present free-for-all in claims continues, functional products will remain, in consumers' minds at least, confusing and controversial, with their contribution to public health correspondingly diminished.

5 REFERENCES

1   National Food Alliance, *Recommendations to the Food Advisory Committee on Functional Foods and Health Claims*, Appendix 5, *Misleading Claims and Functional Foods*, NFA, London, 1996, plus subsequent monitoring by the

author of decisions in the Advertising Standards Authority Monthly Report, London.

2   Ministry of Agriculture, Fisheries and Food, *Food Advisory Committee Review of Functional Foods and Health Claims*, 16 December 1996, London.

3   Biotechnology and Biological Sciences Research Council, *Food Directorate Strategy*, Swindon, 1997.

4   R. Righelato, *Functional Foods & Health Claims of Interest, Feedback*, No 18, Winter 1995.

5   Committee on Medical Aspects of Food Policy, *Dietary Reference Values for Food Energy and Nutrients for the United Kingdom*, para 1.3.3., HMSO, London, 1991.

6   R. Whitehead, cited in Feedback, op cit.

7   K. L. Wrick, *The potential role of functional foods in medicine and public health*, in I. Goldberg (Ed), *Functional Foods*, Chapman & Hall, New York, 1994.

8   R. I. San Lin, *Phytochemicals and Antioxidants*, in Goldberg, op. cit.

9   American Soybean Assn, Second International Symposium on the Role of Soy in Preventing and Treating Chronic Disease, 1996.

10   I. Goldberg, op. cit.

11   M. Ashwell (Ed), *Tomorrow's Nutrition*, Nutrition Bulletin 1992 (suppl), British Nutrition Foundation, London.

12   J. T. Winkler, *Functional Foods: The Challenges for Consumer Policy*, Consumer Policy Review, Vol 6, No 6, Nov/Dec 1996

13   J. Basu, *Genetic Roulette*, Stanford Today, Nov/Dec 1996.

14   R. Bird, *This is the Rest of Your Life*, Guardian, 7 January 1997

15   J. T. Winkler, *From Functional Foods to Patent Medicines*, Low & Light Digest, September 1996.

16   American Soybean Association, op. cit., and EU FAIR Programme Research Abstract 0085, March 1995.

17   *The Food Labelling Regulations* 1996, SI 1996/1499, HMSO. Part III and Schedule 6.

18   Department of Health, *Folic Acid and the Prevention of Neural Tube Defects:* Report from an Expert Advisory Group, London, 1992.

19   R. Weindruch and R.L.Walford, *The Retardation of Aging and Disease by Dietary Retriction*, Thomas, Springfield, Ill; 1988.

20   E. J. Masoro, *Food Restriction Research: Its Significance for Human Aging*, American Journal of Human Biology, 1, pp 339-45, 1989.

21   Department of Health, *The Health of the Nation: A Strategy for Health in England*, HMSO, London, 1992.

22   Department of Health, *Eat Well; Eat Well II*, London, 1992 and 1994.

23   H. R. Warner and S. K. Kim, *Dietary Factors Modulating the Rate of Aging*, in Goldberg, op. cit.

24   J. S. Bland and D. G. Medcalf, *Future Prospects for Functional Foods*, in Goldberg, op cit.

25   A. Drenowski, cited in *Emerging Research on DHA and the Brain*. Reported at US Conference, Nutraceuticals International, Vol 2, No 6, June 1997.

26   Personal communication from Campbell Soups.

27   EU, Commission Directive 91/321/EEC of 14 May 1991 (OJ No L175, 4.7.91);

Department of Health, *The Infant Formula and Follow-On Formula Regulations* 1995, SI 1995/77.

28   Three times in the past three years complaints against memory enhancement claims by gingko products have been upheld by the ASA. In all three cases, the issue has been whether the available evidence on older subjects could be extended to other age groups.

29   MAFF, *Annual Report 1995 of the Advisory Committee on Novel Foods and Processes*, para 2.2 on long-chain polyunsaturated fatty acids (LCPs).

30   Anon, *Vitamin E Shows Benefit in Alzheimer's, But More Study Needed*, Nutraceuticals International, Vol 2, No 5, May 1997.

31   H. L. Meiselman and H. R. Lieberman, *Mood and Performance Foods*, in Goldberg, op. cit.

32   Independent Television Commission, *ITC Television Advertising Complaints Report*, May 1992.

33   Office of Science & Technology, LINK Agro-Food Quality Programme, Project AFQ39, *Dietary and Sensory Effects on Mood and Cognitive Efficiency: Relevance to the Acceptability and Choice of Foods and Beverages*, 1994-97.

34   Anon, *News Reports on Regulatory Action*, Nutraceuticals International, June and October 1996.

35   Regulation (EC) No 258/97 of the European Parliament and of the Council, concerning Novel Foods and Novel Food Ingredients, Official Journal of the European Communities, 14.2.97, No L 43/1.

36   A. L. Cochrane, *Effectiveness and Efficiency*, Nuffield Provincial Hospitals Trust, London, 1972.

37   National Consumer Council, *Messages on Food*, London, 1997.

38   National Food Authority, *Review of the Food Standards Code: Concept Paper on Health and Related Claims*, Canberra, Australia, 1996.

SCIENTIFIC AND REGULATORY ISSUES ABOUT FOODS WHICH CLAIM TO HAVE A POSITIVE EFFECT ON HEALTH

D.P. Richardson

Nestlé UK Ltd
St George's House
Croydon, Surrey
CR9 INR

1 INTRODUCTION

A successful food industry depends not only upon confidence that the food product is safe but also upon the ability to innovate and to meet and satisfy consumer requirements. Pressure to innovate is customer driven and failure to respond will mean lost profits and lost customers. In the last 10 years, there are many examples which testify to innovation in the food and drink industry and arguably manufacturers and retailers are more innovative now than ever before.

For example, we can witness a whole range of packaging technologies—the indispensable tools for the distribution of safe and healthy foods; the increased commitment to quality in the spread of standards such as 1S0/9000; the advent of the information revolution and the streamlining of logistical systems and trade channels between farmers, manufacturers and retailers; and the innovative ways to communicate with consumers through product information such as access to databases and phone/care-line services. However, the biggest revolutions continue to be seen in the increasing variety of products, the improvements in ingredient technology, the spread of international tastes and the drive to create and sell foods and drinks that offer consumers nutritional and health benefits.

2 NUTRITION SCIENCE AND PRODUCT DEVELOPMENT

The incorporation of nutrition research efforts and thinking into product design has a long and respectable history. Products conceived with micronutrients in mind are well established in the range of everyday shopping items, for example breakfast cereals with several B vitamins and minerals; margarines and spreads with vitamins A, D and E; soft drinks with vitamin C; iodised salt etc. These fortified foods and drinks, and indeed nutrient supplements, originate from research on individual nutrients going back to the beginning of the 20th Century.

The principle of adding micro quantities of active ingredients to a component of the food supply to achieve a health benefit is well demonstrated by the addition of fluoride to drinking water to delay the onset of dental caries. Several nutrients are added to food and drink products around the world as public health measures and as cost-effective ways of ensuring the nutritional quality of the food supply.[1] Additions of nutrients have also formed the basis of several marketing strategies in new product development.

More recently, products with lower fat and saturated fatty acids and higher contents of different fibres have been successful in the marketplace, and have their origins in the disease risk-management phase of nutrition research eg diets low in fat and saturated fatty acids to reduce risk of heart disease and some cancers. The current national and international scientific investigations linking the role of diet to disease prevention and optimisation of human performance has created further opportunities for product development especially with the increased public interest in physical fitness and overall physical and mental wellbeing.[2]

3 SATISFYING CUSTOMER DEMAND

Market research continues to show that the consumer appetite for healthy products, for whatever reason, remains insatiable. However, public concerns and interest in food safety and nutrition have propelled food science, technology and nutrition out of the laboratory and firmly onto the nation's dinner plate. New foods and ingredients already face formidable barriers to acceptance and there will be even more resistance to products which are perceived to originate in the laboratory rather than in the kitchen. Hence, the development of some functional foods and the success of products of biotechnology will be dominated by their consumer acceptability, and marketeers and technologists will need to take care to communicate the best image to prevent misunderstandings or misrepresentations.

Surveys in the UK continue to indicate that far more people are concerned about food and their health than in the past. The National Health Survey of over 900 women aged 16 to 64 years provides a continuous assessment of consumer awareness of and attitudes towards health matters relating to the purchase and consumption of food and drink, including their spontaneous responses to what they consider to be the attributes of 'unhealthy' and 'healthy' foods.[3] These consumer perceptions towards food components are shown in rank order in Figures 1 and 2. Table 1 highlights the trends over the last eight years which have provided many opportunities for product innovation and new product developments.[3] Similarly, the consumer and market studies carried out by the Leatherhead Food Research Association provide evidence of increasing awareness and knowledge of the foods and ingredients which may be beneficial to health and the prevention of disease.

4 POSITIVE EATING

Whereas many of the perceptions on healthy eating focus on avoidance of certain

Any fat	86%
Fat in general	79%
Too much/added sugar	48%
(Too much added) salt	22%
Additives	20%
E numbers	13%
Animal fats	10%
Cholesterol	8%
Preservatives	8%
Saturated fats	7%

Figure 1 *Spontaneous consumer perceptions of contents of unhealthy food components* (n=909:1996 National Health Survey)

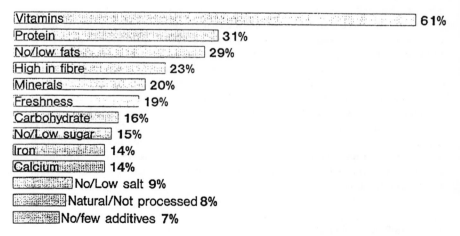

Vitamins	61%
Protein	31%
No/low fats	29%
High in fibre	23%
Minerals	20%
Freshness	19%
Carbohydrate	16%
No/Low sugar	15%
Iron	14%
Calcium	14%
No/Low salt	9%
Natural/Not processed	8%
No/few additives	7%

Figure 2 *Spontaneous consumer perceptions of contents of healthy food components* (N=909:1996 National Health Survey)

Table 1 *Contents of 'Healthy' and 'Unhealthy' Foods; Spontaneous %, National Health Survey*

	1988	1989	1990	1991	1992	1993	1994	1995	1996
Vitamins	36	47	56	56	57	56	57	58	61
Protein	28	33	31	31	33	34	34	34	31
Fibre/ Roughage	31	33	34	36	32	30	30	30	23
Low fat	19	25	21	25	28	23	26	28	29
Minerals	11	18	17	15	19	16	19	23	20
Fat (any mention)	57	65	74	76	81	83	84	84	86
Fat (in general)	47	56	66	66	72	74	75	77	79
Sugar (any mention)	45	51	49	51	50	51	50	49	48
Salt (too much)	21	19	20	26	25	22	21	19	22
Additives	32	33	29	29	24	19	21	16	20

components in foods, there is an important and gradual change in the nature of the healthy eating market towards more positive messages. There is an increasing recognition that people can help themselves and their families to prevent future illness and disease and to maintain their state of health and wellbeing through informed dietary practices. For example, the important role of fruit and vegetables in disease prevention including some cancers and the latest research on dietary antioxidants and combinations of phytoprotective substances in plants will undoubtedly provide the necessary impetus to the further development of positive nutrition in Europe.

5 PREVENTION IS BETTER THAN CURE: GOVERNMENT STRATEGIES FOR HEALTH

Ageing populations, the recognition by government of the economic significance of illness prevention and the increasing evidence that preventive measures should be implemented early in life have stimulated many governments to set in place comprehensive strategies for promoting health across the whole community. In the UK, the publication of the Health of the Nation White Paper,[4] the establishment of the Nutrition Task Force (NTF) and its Project Teams provided the opportunity for a co-ordinated programme of action to implement the nutritional aspects of the Government's health strategy.[5]

The NTF drew together a wide range of expertise from food manufacturers and retailers, caterers, health professionals, consumers and the voluntary sectors to work with Government Departments. Coronary heart disease (CHD) and strokes, cancers and obesity were among those areas which are recognised as being major causes of premature death or avoidable ill health and where effective interventions should be possible, offering significant scope for improvements in health.

In a recent article, Richardson focused on the positive contributions of the food and drink industry to health promotion and policy making in the context of the Health of the Nation initiatives and the development of effective nutrition policies for the UK.[6] To date, the responses in new product developments have demonstrated the willingness and technical expertise of the food industry to provide a greater choice and variety of foods consistent with the Health of the Nation's targets for a gradual downward shift in fat intake.

The NTF also recognised that if the targets for the reduction of fat in the diet are to be met it is necessary for people to increase their consumption of carbohydrate-containing foods, mainly bread, pasta, rice and other cereal products together with more vegetables, fruit and oily fish. As previously mentioned, much of the current dietary advice is negative in tone. Too little emphasis has been placed on the more desirable things to eat which everyone can consume in generous amounts. For example, the positive message of special importance is the recommendation to eat at least five servings a day of fruit and vegetables, fresh and preserved, not least because these foods are the principle sources of the major antioxidant nutrients vitamin C and vitamin E, but also they contain a whole host of antioxidant and phytoprotective substances like beta-carotene in carrots, lutein in broccoli, lycopene in tomatoes and flavonoids in tea, apples and onions. These substances and many more are believed to exert their effects individually and synergistically by counteracting oxidative processes that contribute to the development of chronic diseases.[7,8]

6 FOODS WHICH CLAIM TO HAVE A POSITIVE EFFECT ON HEALTH

Poor food choices and restricted diets can effect nutritional status of individuals at any stage of life and potentially their long term health. Hence, there are many opportunities for the food industry to manufacture specially designed, added-value products to meet the nutritional requirements of young children, adolescents, women of childbearing age, sports people, the middle-aged and older people. Although confusion still exists about how best to define these evolving areas of food and ingredient technology, the common thread is that the foods or components found within them have a potential beneficial role in positive nutrition and in the reduction of risk of disease.

Food manufacturers see an opportunity to develop products which contain beneficial components as part of a normal, varied diet, based on the principles of good nutrition, eg a wide variety of foods, balance and moderation. Key areas of research interest include antioxidant substances including beta-carotene, vitamins C and E (the 'ACE' nutrients), mineral nutrients such as calcium, magnesium, zinc and selenium, disease preventive phytochemicals such as the flavonoids, probiotics like Lactobacillus acidophilus 1,[9] fatty acids and lipids including fish oils, and a range of macromolecules including dietary fibres and oligosaccharides.

7 NUTRITION AND DISEASE PREVENTION

7.1 Folic Acid and Prevention of Neural Tube Defects

There is now compelling evidence that the majority of neural tube defects (NTDs) can be prevented by an adequate intake of periconceptional folic acid.[10] This relationship between a vitamin and reduction of risk of disease is the culmination of over 25 years' research and has highlighted a number of challenges for nutrition policy. These challenges include the ways in which dietary folate intake can be increased, the risks and benefits of food fortification and the opportunity to make health claims on labels of foods and dietary supplements.

NTDs develop very early in pregnancy (18-30 days after conception) and hence folate status is important at the periconceptional period before a woman knows she is pregnant. Practically, good nutrition must be maintained throughout the childbearing period. Folate intake can be increased in three complementary ways: by taking a folic acid supplement; by eating folate-rich foods; and by eating foods fortified with folic acid.[11] Humans obtain most of their folate from fruits and vegetables and typical intakes are 0.15 to 0.2mg/day. Folate intake needs to be increased about threefold to reach the level of 0.4mg/day, the amount of folate known to be effective against NTDs, and which does not mask the diagnosis of pernicious anaemia and/or vitamin B_{12} deficiency.

Many countries have now issued public health advice, recommending a combination of dietary and supplemental means to increase periconceptional folate intake and many have recommended fortification of food with folic acid. In the USA, the Centers for Disease Control and Prevention[12] concluded that the development and implementation of a fortification programme for the addition of folic acid to the food supply could be effective in increasing the folate intake of women of child-bearing age. Because the Food and Drug Administration (FDA) has a mandate to set fortification levels that are safe for all population groups, and the fact that the effects of long-term high intakes of folic acid are not well-known, the FDA ruling has been designed to keep folic acid intake under 1mg/day. Hence, fortification will be mandatory from 1 January 1998, and folic acid must be added to flours, breads, corn meals, rice, noodles, macaroni and other grain products at a level of $140\mu g/100g$. These foods were chosen for folate fortification because they are staple foods and they have a long history of being successful vehicles for improving nutrition by reducing the risk of classic nutrient deficiency diseases. In the UK, efforts are being made to improve further the knowledge about folate and NTD prevention and an increasing number of breads and cereals are being fortified with folic acid. For example, the level of folic acid added to breakfast cereals is between 50 and $100\mu g$ folic acid/serving and a number of soft-grain breads contain $105\mu g$/serving. These additions, but not the levels, are governed by The Food Labelling Regulations 1996. Currently there is discussion on whether to make fortification mandatory and the UK food industry has stated that, if the medical view is that addition of folic acid to flour would be a safe and effective means of reducing the incidence of NTDs, a statutory approach should be followed.

In the UK, food law prohibits claims that a food has the ability to prevent, treat, or cure disease. However, it is possible to draw attention to the beneficial role of nutrients within a food either on pack or in advertising and other promotional activities. These means of communication are effective ways of raising public awareness and supporting other programmes aimed at increasing population folate intake. In the USA, in the Federal Register of October 14,1993, the FDA proposed to authorise a health claim about the relationship between folate and the reduced risk of NTDs on food labels. In the final rules on folic acid fortification, effective 1 January 1998, manufacturers will be allowed to make claims on labels that fortified products contain folic acid and that an adequate intake of the nutrient may reduce the risk NTDs. It should be noted that unenriched cereal-grain products without folic acid added will continue to be available to consumers who do not wish to use them. In conclusion, the additions of folic acid to foods will continue to attract interest from academia, government, consumer and industry sectors with respect to the application of scientific knowledge and the issues of voluntary versus mandatory fortification, bioavailability, level of addition, risks and benefits, health claims and freedom of choice.

Recent research has also shown that an elevated plasma homocysteine concentration is an independent risk factor for cardiovascular disease (CVD), and that such elevations are widespread in the normal population and that dietary folic acid at levels found in existing fortified products such as breakfast cereals have a homocysteine lowering effect. Hence, the benefits of folic acid used in fortified foods to achieve optimal folate status may well have implications for reducing risk of both NTD and CVD.[13]

8 OPTIMISATION OF PHYSICAL AND MENTAL PERFORMANCE

In 1911, when Casimir Funk enriched the language with the new word 'vitamine', a combination of the words 'vital amine', he would have scarcely believed how far the research would develop and the extent to which additions of nutrients to foods would benefit public health around the world. For a variety of reasons, however, many people still do not achieve the RDAs for specific essential micronutrients. In the UK, changing lifestyles, decreasing energy intakes, the existence of vulnerable groups, such as women of child-bearing age, the increasing elderly population, slimmers etc., those who may not be able to afford the variety of foods necessary for a healthy, balanced diet and those who do not have the knowledge to make adequate food choices, are some of the socioeconomic reasons why it is timely to re-evaluate the public health issues associated with the additions of nutrients to foods.[14] Recent advances in nutrition science have provided increasing evidence that intakes of essential micronutrients not only prevent deficiency states but also have the potential to optimise physical and mental performance and reduce risk of chronic disease and disability.

Additions of nutrients to foods provide one of the safest ways of ensuring the nutritional status of populations and individuals because the quantity of a food one has to eat to reach any potentially hazardous level of the few nutrients that are known to be toxic at high level, limits the risk substantially. Generally, there are few significant, safety concerns arising from foods with added nutrients.[14] Intakes are generally well within safety limits and the risk of nutrient overdosing and imbalance is outweighed by the benefits to public health. The UK has a long tradition of adding essential nutrients to foods, and enrichment practices have been done safely and effectively for over 50 years. Enriched foods, which are familiar and enjoyable, already contribute significantly to the UK diet, and the food industry in many parts of the world has demonstrated that it has taken, and will continue to take, a responsible and sensible approach to both statutory and voluntary additions of nutrients to foods.

The recent scientific evidence for the roles of vitamins and minerals (eg antioxidant functions, hormone-like actions, optimisation of the immune function and metabolic controls), indicate that the daily requirements go beyond the levels of intake for prevention of clinical disorders. There are increasing data which show that diets rich in antioxidant micronutrients are associated with lower risk of premature death from CVD and cancer, and governments are being quick to recognise the economic significance of illness and disease prevention.

These recent advances, demonstrating the potential disease-preventing role of certain nutrients at levels well above the recommended daily amounts (RDAs) raise a number of key issues, not least that the additions are safe. Most nutrients are safe at levels used in food enrichment practices, even at high intakes. With responsible fortification procedures, and addition levels usually as fractions of the RDA per serving, it is actually quite difficult, both for technological and sensory reasons, and as food intakes are limited, to exceed the upper safe levels. In those cases of nutrients where there are narrow safety margins, a careful risk/benefit analysis is required.[14] Progress in this area would help develop the new concepts of 'optimal' health, and stimulate further research with the relationships between higher micronutrient intakes and potential reduction of risk of certain diseases.

9 POTENTIAL LEGISLATIVE LIMITATIONS TO THE USE OF HEALTH CLAIMS

It costs money to research, develop and add health-enhancing ingredients and nutrients to foods and the commercial benefits of doing so would soon be lost if regulatory controls were or became too prohibitive. Regulators will need to satisfy the needs of industry for innovation and marketing but protect consumers from false and misleading claims about products.

Whereas current regulatory provisions may well accommodate the topic of functional foods and health claims, even at these early stages of their development, there is a need for researchers, industry and regulators to collaborate on a regular basis so that the principles and criteria for substantiating claims can be assessed properly. It is also extremely important to involve the enforcement authorities and consumer representatives with the aim of establishing more clearly their points of concern and any ideas for action.

Internationally, there are enormous variations in regulatory approach and these have been summarised by the Australian National Food Authority in their Functional Foods Policy Discussion Paper.[15] In the USA, health claims are permitted by the FDA in a total diet context, in several areas where there is '*significant agreement among qualified experts supported by the totality of publicly available evidence*'.[16] The functional foods industry, however, in the USA dislikes the lack of owner specific health claims and the lack of data protection. In Japan, the decision process focuses more on the active constituents in particular foods, but there the licencing procedure and the conditions for approval are believed to be too stringent and costly.[17] Most countries are maintaining a responsible and cautious approach to the issue of health claims and they will learn from the experiences of others.

In the UK and Europe, a formula for allowing certain types of health claims is being sought through ongoing discussions. The authorities appear not to be against claims establishing a link between nutrition and health and wellbeing, but they have not reached a consensus on how to do it. However, some recent health claims on food products in the UK have been the subject of complaints from consumer associations and MAFF has announced that the Food Advisory Committee should review the situation. The Committee on Medical Aspects of Food Policy, an independent group of experts appointed by the UK Department of Health, has also taken initial steps to clarify the situation. Obviously, at the moment, any health claims require careful interpretation of existing national and international rules. Any unscrupulous attempts to bypass existing regulatory constraints would be unwise and would incur the wrath of industry, government and consumer organisations.

The responsibility for a health claim clearly lies with the company involved. It is important for manufacturers and their trade associations to take the necessary initiatives to comply with the legislation and activities of relevant authorities and also to take measures aimed at eliminating consumer problems and flaws in the market that could undermine credibility. For example, in Sweden, the Federation of Swedish Food Industries, in consultation with the National Food Administration and the National Board for Consumer Policies, prepared a set of 'Food Industry Rules' for the use of health claims in the marketing of food products.[18] Industry took the lead by demonstrating a commitment to the promotion of responsible marketing and advertising (including labels, leaflets, recipes, brochures, videos, films, etc.) and a willingness to accept the burden of responsibility regarding the use of health claims.

The rules are based on state-of-the-art research findings, and they are intended to be updated to keep pace with new developments. The implications for international harmonisation especially are unclear if individual countries are to proceed to regulate with completely different approaches. However, efficient, innovative and profitable investment by the food industry requires a sound regulatory framework, international markets and free movement of goods.

10 SCIENTIFIC CONSENSUS AND THE SUBSTANTIATION OF HEALTH CLAIMS

A major challenge for those involved in the research and development of foods which claim to have a positive effect on health is the scientific validation and substantiation of a claim in the eyes of the law. It is already clear that in some areas, manufacturers will need better clinical evidence of the overall relationship between diet and disease and they may need to carry out specific clinical trials on their products. The issue of substantiation of claims covers not only the safety and efficacy of the food component(s) themselves, but also the finished food as it would be used by people.

Like dietary recommendations, guidelines and quantified targets for intakes of micro- and macronutrients, the substantiation of health claims must be based on sound science —a synthesis of the existing literature, epidemiological, metabolic, animal studies, human clinical evaluations and mechanistic data. The diet and health relationship is complex, data is often difficult to interpret and hence, there is a need to move prudently and conservatively to evaluate the results of nutrition research. In some areas of nutrition, scientific consensus does not exist, and some of the confusing nutritional advice in the public domain reflects conflicting beliefs of medical, scientific and nutritional experts, the effects of media advertising and even political expediency.

The food industry recognises that the greater use of health messages must be consistent with substantiated medical fact and best practice around the world. However, what could be extremely frustrating to reputable manufacturers is that functional foods backed by considerable research efforts and investment could be undermined by, and appear to be the same as, those that are crude, carelessly made, lacking substantiating evidence, and, at worse, fraudulent. If the consumer did not believe or trust the ability of a product to provide the stated benefits, the long-term credibility of the industry would soon be damaged.

The process for the establishment, verification and use of claims and messages must, therefore, be scientifically sound and credible. At the same time, the process must be flexible and pragmatic. It must evolve responsibly with a consensus of regulatory, academic, industry and consumer bodies rather than as a system of rigid product registration, closed lists of authorised claims, predefined wording and regulatory constraints. Table 2 suggests a list of the main areas of focus for substantiation of the nutritional safety and efficacy of functional foods and ingredients that may or may not be defined as novel. It can serve as a useful guide to manufacturers on how to prepare a comprehensive scientific dossier to support health claims. [2]

Table 2 *Basis for the Scientific Dossier to Substantiate Nutritional Safety*

• Nutritional composition
• Dietary significance; intake; extent of use
• Interactions with other components of diet; bioavailability
• Presence of antinutritional factors
• Implications for possible changes in gut microflora
• Quantitative effects; dose response
• Impact on metabolic pathways and physiological function in humans
• Overall toxicological assessments including allergy/intolerance factors
• Potential effects on vulnerable groups, young, elderly etc
• History of safe use; previous human exposure
• Storage, preparation and instructions for use
• Direct effects on pathophysiological processes
• Relation to current dietary recommendations/targets
• Technical details of processing and product specification
• If probiotic, history of organism, consistency and stability of organism, survivability, colonisation, replication, amplification in human gut

11 RESEARCH AND DEVELOPMENT

The number of major research programmes designed to investigate and clarify the therapeutic value of foods and food components are forecast to continue to grow, particularly where serious debilitating diseases are concerned, eg CHD, cancers, osteoporosis etc. This growth is happening in both the private and public sectors and we are seeing converging technologies and scientific disciplines throughout the food supply chain as well as in basic and applied research respectively. There are also increasing opportunities for partnerships and collaborative efforts between researchers, government and industry. Table 3 summarises the main areas for further work in this area of scientific, technical, regulatory and consumer interest.

In the UK, government departments, the research councils and the food industry, both individual companies and through their Trade Associations, have participated fully in the development of national strategies for establishing a healthy diet leading to an improved quality of life. In addition to fundamental issues relating to the quality of life, it is emphasised that there is the potential for wealth creation through the exploitation of such research by the food industry. The findings of the UK Technology Foresight Programme[19] (Food and Drink and the Health and Life Sciences Panels) also identified the relationship between food and health as a significant factor in the quality of life and future competitiveness of the UK. The main objectives of these new research initiatives include:

- increasing understanding of the mechanisms underlying the physiological, metabolic and behavioural effects of specific components of the diet from molecular to the whole organism level
- linking food and plant science with nutrition research
- taking advantage of the opportunities of the European nutrition research programme and supporting complementary research in the UK
- encouraging technology interaction through LINK and by promoting dissemination of research findings to industry and the consumer.

MAFF have sponsored an Agro Food Quality Link Programme Functional Foods Initiative in which over one hundred companies from all sectors of the food industry have taken part. A series of focus groups are concentrating on specific topics and LINK projects have already been approved, including one on methodologies for measuring antioxidant status and indices of oxidative stress in man.[20]

12 CONCLUSIONS

The food industry has created efficient and safe food processing and delivery systems around the world, and it is willing and able to capitalise on new technologies and research with the development of tasty, pleasurable and healthy products. The development of foods of the future requires careful attention to safety, labelling and claims, as well as to the nutritional and physiological rationale, cost and sensory

Table 3 *Research and Development Needs*

• Investigation of health-enhancing diets, foods, ingredients and nutrients for: Safety
Dose-response
Efficacy over time
Public health
• Substantiation of potential health benefit claims
• Communication of health benefits, appropriate labelling and marketing
• Pioneering and innovative technological advances in the food industry
• Appropriate and secure regulatory framework to commit resources

qualities, all of which are keys to food choice and acceptability. Developments in the area of functional foods will be heavily dependant on scientific substantiation of which safety and efficacy will be the principle components.

Foods that claim to have a positive effect on health represent one of the most challenging areas of food technology and nutrition science and there is a need to create a regulatory environment that protects the consumer, permits fair trade and provides the necessary controls to ensure validity of claims. An appropriate and secure regulatory framework is necessary to commit resources to fundamental nutrition research and for the application of science in pioneering and innovative product developments. The Food Advisory Committee review of functional foods and health claims has provided the opportunity for constructive dialogue between regulators, academia, industry and consumer representatives in order to develop guidelines which food manufacturers and retailers should follow when making health claims on their products. Opportunities now exist in both generic areas of science and new areas of scientific development which could be of equal benefit to the consumer.

The pace of innovation, however, is dependent on the economic state of the organisations and institutions paying the bills and the level of imaginative research funding to match an area rich in ideas. The creation of a regulatory climate which encourages fair trade, free movement of goods, the harmonisation of controls to ensure validity of claims, and consumer protection are fundamental to the future role of functional foods for consumers and food markets. Ultimately, however, we should always remember that enjoyable eating and drinking involves balance, variety and moderation within the framework of a healthy, active lifestyle.

13 REFERENCES

1. D.P. Richardson, Food Fortification. In: Technology of Vitamins in Food. Edited by P. Berry Ottaway. Blackie Academic and Professional, Glasgow 1993 pp 233 - 245.
2. D.P. Richardson, Functional Foods - Shades of Grey : An industry perspective. *Nutrition Reviews* 1996, 54 (11), S174 - S185.
3. JRA Research, National Health Survey, Halifax House, Halifax Place, Nottingham, NGI IQN. 1995.

4. Secretary of State for Health. Health of the Nation : A strategy for health in England, HMSO, London. 1992.

5. Department of Health, Eat Well: An action plan from the Nutrition Task Force to achieve the Health of the Nation targets on diet and nutrition. Department of Health. March 1994.

6. D.P. Richardson, UK Food industry responses to the Health of the Nation White Paper *Brit. Food J.*1995, 97(2), 3 - 9.

7. G. Block, B. Patterson, and A. Suber, Fruit, vegetables and cancer prevention a review of epidemiological evidence. *Nutrition and Cancer* 1992, 18:1 - 29.

8. The American Dietetic Association : Position paper on Phytochemicals and functional foods. *Journal of the Amer. Dietetic Assoc.*1995, 95(4):493 - 496.

9. D.P. Richardson, Probiotics and product innovation. *Nutrition and Food Science* 1996, (4), 27 - 33.

10.L.E. Daly, P.N. Kirke, A. Molloy, D.G. Weir, and J.M. Scott, Folate levels and neural tube defects. Implications for prevention. *Journal of the American Medical Association* 1995, 274,1698 - 1702.

11. C. Bower, Issues in the prevention of spina bifida. *J. Royal Society of Medicine* 1996, 89, 436 - 442.

12. Centers for Disease Control and Prevention. Recommendations for the use of folic acid to reduce the number of cases of spina bifida and other neural tube defects. Morb. Mortal Wkly report 1992, 41, 1 - 7.

13. J.M. Scott and D.G. Weir, Homocysteine and cardiovascular disease. *Quarterly Journal of Medicine* 1996, 89, 561 - 563.

14. D.P.Richardson. Additions of nutrients to foods. (In Press) Proceedings of the Nutrition Society 1997 (presented on 19th Feb.1997 Nut. Soc. Winter Meeting)

15. Australian National Food Authority. Functional food policy discussion paper.Canberra: Australian National Food Authority, 1994.

16. Nutrition Labelling : final rules. Fed. Regist 1993; 58(3).

17.T. Furukawa, The nutraceutical rules : health and medical claims : "Foods for specified health use" (FOSHU) in Japan, Regulatory Affairs 5,189 - 202. 1993.

18. Federation of Swedish Food Industries: Health claims in the marketing of food products. The Food Industry's Rules. Sveriges Livsmedelsindustri forbund August. 1990.

19. Cabinet Office. Office of Public Service and Science. Office of Science and Technology. Forward Look of Government-funded Science, Engineering and Technology, London, HMSO. 1995.

20. R.A. Riemersma, LINK Agro Food Quality Programme. Dietary antioxidant vitamins and oxidative stress. Cardiovascular Research Unit, University of Edinburgh in collaboration with Nestle' UK Ltd; Van den Bergh & Jurgens Ltd and Roche Products Ltd. 1995.

CONCLUSIONS

M.J. Sadler

Senior Nutrition Scientist
British Nutrition Foundation
52-54 High Holborn
London, WC1V 6RQ

It is the British Nutrition Foundation's role to encourage the impartial and independent interpretation of nutritional scientific knowledge. A particularly satisfying outcome of the conference was the achievement of our main aim, and that of the Royal Society of Chemistry—to *critically* review the area of functional foods and health claims, and to provide a forum for frank and open discussion.

Though the concept of functional foods is not a new one, it is the development of functional health claims that is a recent and taxing issue. The regulatory environment concerning this development will be strongly influential in determining the future of functional foods.

One of the notable outcomes of the conference was the integrated nature of the meeting and the way the different sections gelled together. For those who had, in some cases the luxury, and in others the insight to attend the whole conference, it was an opportunity to gain an understanding of the full picture—from the area of gathering scientific evidence for claims, ensuring the products do what they claim, and assessing the evidence in order to make a claim. Having an appreciation of the background and potential problems in each of these areas is a useful knowledge base from which to ensure informed contributions to the debate.

The conference highlighted the growing scientific evidence for health enhancing dietary components. However a number of valid concerns were also raised—for example the need to assess risks against benefits, the potential problems of overconsumption, and the need to consider potential dietary components as part of a matrix in the context of the whole diet and not in isolation.

One of the conclusions that came across strongly was the need to take a long-term view rather than a short-term approach—with less emphasis on a marketing drive and more emphasis on a strong scientific base to developments. The need to ensure adequate scientific evidence to substantiate health claims is paramount. Discussions of how much evidence and what types of studies are needed was an interesting aspect of the conference, and will undoubtedly be the subject of an on-going debate.

We know that the general population is remarkably resistant to dietary advice. The availability of products to help people improve the nutritional balance and health promoting effects of their diet is a positive concept that should be encouraged as one way forwards. The experience from some dietary intervention trials is that the provision of appropriate modified foods can aid subjects in achieving the desired dietary manipulations and compliance. There clearly is an opportunity for functional foods to play a role in improving the nation's diet as well as helping people in specific target groups.

The functional food concept and the development of health claims regulations are still in their early stages. This conference made a valuable contribution in helping to ensure the area moves forwards in a positive direction.

Subject Index